MICROSCOPY HANDBOOKS 45

Immunoenzyme Multiple Staining Methods

Royal Microscopical Society MICROSCOPY HANDBOOKS

Series Advisors

Angela Kohler (Life Sciences), *Biologische Anstalt Helgoland, Notke-Strasse 31, 22607 Hamburg, Germany*

Mark Rainforth (Materials Sciences), *Department of Engineering Materials, University of Sheffield, Sheffield S1 3JD, UK*

Immunoenzyme Multiple Staining Methods

C.M. van der Loos

Department of Cardiovascular Pathology, Academic Medical Centre, Amsterdam, The Netherlands

In association with the Royal Microscopical Society

© BIOS Scientific Publishers Limited, 1999

First published 1999

A CIP catalogue record for this book is available from the British Library.

ISBN 1 85996 187 8

BIOS Scientific Publishers Ltd
9 Newtec Place, Magdalen Road, Oxford OX4 1RE, UK
Tel. +44 (0)1865 726286. Fax +44 (0)1865 246823
World Wide Web home page: http://www/bios.co.uk/

Published in the United States of America, its dependent territories and Canada by Springer-Verlag New York Inc., 175 Fifth Avenue, New York, NY 10010-7858, in association with BIOS Scientific Publishers Ltd

Production Editor: Andrea Bosher.
Typeset by Marksbury Multimedia Ltd, Midsomer Norton, Bath, UK.
Printed by The Cromwell Press, Trowbridge, UK

Front cover: Triple-staining technique according to the protocol in *Appendix H.1.3 (see colour plate 13).*

Contents

Colour plates can be found on pages 16–31

Abbreviations

AEC	3-amino-9-ethylcarbazole
AP	alkaline phosphatase
APAAP	alkaline phosphatase anti-alkaline phosphatase
BCIP	5-bromo-4-chloro-3-indolyl-phosphate
Bluo-GAL	5-bromo-3-indolyl-β-D-galactoside
DAB	3,3$'$-diaminobenzidin
DIG	digoxigenin
DNA	deoxyribonucleic acid
EDTA	ethylenedinitrilo tetraacetic acid
EPOS	enhanced polymer one-step staining
FITC	fluorescein isothiocyanate
GAL	β-galactosidase
GAM	goat anti-mouse immunoglobulin
GAR	goat anti-rabbit immunoglobulin
HRP	horseradish peroxidase
Ig	immunoglobulin
IGS	immunogold/silver
NBT	nitro blue tetrazolium
PAP	peroxidase anti-peroxidase
PBS	phosphate-buffered saline
RAM	rabbit anti-mouse immunoglobulin
RNA	ribonucleic acid
SAG	swine anti-goat immunoglobulin
SAR	swine anti-rabbit immunoglobulin
Strep	streptavidin
StrepABC	streptavidin–biotin complex
TBS	Tris–HCl buffered saline
TMB	tetramethylbenzidine
Tris	Tris hydroxymethyl amino methane
TUNEL	terminal deoxytransferase (TdT)-mediated dUTP nick end labelling
X-GAL	5-bromo-4-chloro-3-indolyl-β-D-galactoside

Key to figures 3.1–3.9

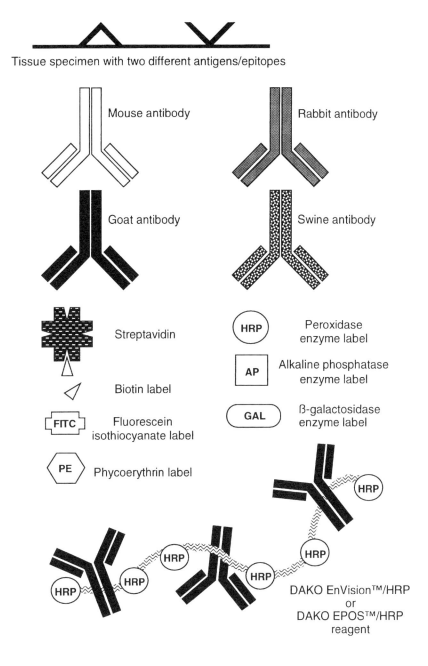

Tissue specimen with two different antigens/epitopes

Mouse antibody

Rabbit antibody

Goat antibody

Swine antibody

Streptavidin

HRP — Peroxidase enzyme label

Biotin label

AP — Alkaline phosphatase enzyme label

FITC — Fluorescein isothiocyanate label

GAL — ß-galactosidase enzyme label

PE — Phycoerythrin label

HRP
HRP
HRP
HRP
HRP
HRP

DAKO EnVision™/HRP
or
DAKO EPOS™/HRP
reagent

Sponsors

I am grateful to the following companies and institutes for their financial support:

Aurion BV, Wageningen, The Netherlands

Cardiovascular Pathology (non-profit) Research Fund, Amsterdam, The Netherlands

DAKO A/S, Glostrup, Denmark

Pediatric Gastroenterology and Nutrician Research Group, Sophia Children's Hospital, Rotterdam, The Netherlands

Institute of Human Nutrition and Department of Pathology, University of Southampton, UK

Southern Biotechnology Associates Inc., Birmingham AL, USA

Training centre (TMB), Hogeschool Brabant, University of Professional Education, Faculty of Technology and Science, Etten-Leur, The Netherlands

Wound Healing Research Group, Academical Medical Centre, Amsterdam, The Netherlands

Preface

Today, histopathological investigations are unthinkable without immunostaining techniques. The technique has come a long way. When immunostaining on histological sections was first introduced, the method was limited technically and the antibodies used were directly conjugated with fluorochromes in order to visualize their site of attachment within the tissues. These techniques revolutionized the diagnostic arena of the pathologist, particularly at first in the fields of renal and skin diseases, but later in many other applications. Up to now, fluorescent techniques for diagnostic and research purposes have remained *en vogue*. However, for many applications, including diagnostic pathology, there are disadvantages inherent in fluorescence microscopy. This has led to a movement to investigate the possibilities of using immunostaining techniques on frozen and routinely formalin-fixed and paraffin-embedded tissue specimens fit for ordinary light-microscopy. After all, this is the method of study most familiar to the pathologist. The search for further improvement and refinement soon opened up new frontiers.

Unlike double immunofluorescence and double immunoelectron microscopy, multiple immunoenzyme staining is mostly covered within a few handbook pages or review paragraphs. Therefore, it was decided to focus this handbook entirely on the application of enzyme tracers and gold particles for light-microscopy in multiple immunohistochemical techniques.

Three manuscripts formed the solid basis of this handbook.

First, my PhD thesis devoted entirely to multiple staining techniques, and still in use as a handbook at many institutes (Van der Loos, 1992). The original introduction to immunohistochemical techniques can be found updated in *Chapter 1*. The thesis contained several papers, which are now summarized in *Chapter 3* describing diverse double-staining concepts.

Second, a double-staining brochure entitled: Immunoenzymatic double staining methods. A practical guide' (Van der Loos, 1998). This practical guide, written on invitation for DAKO A/S (Glostrup, Denmark), is related specifically to their reagents. I am grateful to Mrs Karina Norring Hjort and colleagues, for their valuable contributions, and their permission to use parts of the brochure in the present handbook.

The third manuscript important for this handbook was an updated theory manual which I utilized during the conduction of 10 subsequent editions of the practical course: 'Multiple staining in immunohistochemistry' organized by the training centre (TMB), Hogeschool Brabant, University of Professional Education, Faculty of Technology and Science, Etten-Leur, The Netherlands (website: http://tmb.ftn.hsbrabant.nl/ and e-mail: training.centre@ftn.hsbrabant.nl). I am grateful to the training centre for incorporating this book into future multiple staining courses.

This handbook reflects the development of multiple immunoenzyme staining at the Department of Cardiovascular Pathology in the Academic Medical Centre, University of Amsterdam. Our laboratory regularly received requests for detailed multiple-staining protocols and assistance, mainly because of the good quality of the multiple-staining slides as presented by Professor Anton Becker, head of the Cardiovascular Pathology Department. These requests forced me to create versatile protocols inherent in the 'good laboratory practice' aspects with regard to general applicability and reproducibility. I am grateful to Anton Becker for his support and for encouraging the development of new staining procedures.

The performance of a multiple immunoenzyme staining experiment has all too often been considered as close to a 'magician's act', suggesting that these techniques will only work properly in the hands of a few 'lucky' scientists or technicians. The present book shows that there is a scientific footing to the magician's act and that the practical aspects for optimal results can be regulated. Another more flattering remark concerning double staining was made by a pathologist, who compared the development of single staining to double staining with mono- and stereo sound music, referring to the fact that double staining provides important extra information. Both metaphors, however, refer strongly to the idea that immunoenzyme multiple staining is a real 'speciality', rather than a commonly used technique. Hopefully, this book will contribute to the more frequent use of this colourful methodology by a broader public.

Chris van der Loos

Acknowledgements

The author wishes to express his gratitude to Professor Anton E. Becker, MD PhD (Dept. of Cardiovascular Pathology, Academical Medical Centre, Amsterdam, The Netherlands), Heike Göbel, MD (Dept. of Cardiology/Nephrology, University of Frankfurt/Main, Germany), Jennifer Garry, BSc (Nutrition Department, University of Southampton, UK) and Frank Thies, PhD (Nutrition Department, University of Southampton, UK) for critical reading of this manuscript.

Furthermore, I would like to thank my collegues at the Cardiovascular Pathology Department for all kind of assistance: Allard van der Wal MD PhD, Onno de Boer PhD, Peter Teeling BSc, Mischa Houtkamp BSc, Hanneke Ploegmakers, Marsha Schenker, Petra Zureineh, Peter van Rooij and Wilfried Meun.

Safety

Attention to safety aspects is an integral part of all laboratory procedures and both the Health and Safety at Work Act and the COSHH regulations impose legal requirements on those persons planning or carrying out such procedures.

In this and other Handbooks every effort has been made to ensure that the recipes, formulae and practical procedures are accurate and safe. However, it remains the responsibility of the reader to ensure that the procedures which are followed are carried out in a safe manner and that all necessary COSHH requirements have been looked-up and implemented. Any specific safety instructions relating to items of laboratory equipment must also be followed.

1 Immunohistochemistry: an historical overview

The identification of tissue antigens using a specific antigen–antibody interaction and subsequent microscopical visualization methods, is named *immunohistochemistry*, or when dealing with intact cells *immunocytochemistry*. The technique was firstly introduced as a fluorescence method by Coons *et al.* (1941) and Coons and Kaplan (1950). In 1967 immunoenzyme histochemistry was founded, introducing antibodies conjugated with horseradish peroxidase. This enzyme was capable of generating an insoluble coloured reaction product suitable for light-microscopic visualization (Nakane and Pierce, 1967). In the same paper these authors suggested employing the electron density of the peroxidase reaction product to trace the antibody-binding site at an ultra-structural level. A few years later colloidal gold-labelled antibodies were suggested for this purpose (Faulk and Taylor, 1971).

Until 1980 the applicability of immunohistochemistry was still hampered by the limited number of traditionally raised polyclonal antisera. However, the introduction of hybridoma technology for the production of mouse or rat monoclonal antibodies against a wide range of human cellular epitopes should be acknowledged as the first major breakthrough of immunohistochemistry (Köhler and Milstein, 1975). Since the early 1980s the number of applications of immunohistochemistry has increased drastically, and the technique has established itself as an important medical–biological investigatory tool.

1.1 Tissue-fixation and pretreatment techniques

The immunohistochemical staining technique can be performed on a variety of specimens, such as tissue sections, cut from frozen, paraffin- or resin embedded-blocks, cell smears, cell cultures and cytospin preparations (Polak and Van Noorden, 1997; Taylor, 1980; Van Noorden, 1986). All tissue specimens require fixation prior to the application of immunoreagents. Already at the beginning of immuno-histochemistry, while still dealing with traditionally raised polyclonal antibodies, it was observed that tissue fixatives, particularly the type and duration of fixation, affect the results of immunohistochemical staining (Brandtzaeg, 1982; Curran and Jones, 1977; DeLellis *et al.*, 1979; Miller, 1972; Taylor, 1978). Nevertheless, at that time many polyclonal antibodies could be applied successfully to routinely formalin-fixed and paraffin-embedded tissue sections (Mepham, 1982; Taylor, 1974; Taylor and Burns, 1974; Taylor and Mason, 1974). The use of proteolytic enzymes on formalin-fixed and paraffin-embedded tissue sections, prior to the immunoreagent steps, considerably improved the detection of some antigens (Denk *et al.*, 1977; Mepham *et al.*, 1979). In comparison to the polyclonal antibodies, however, the use of monoclonal antibodies for paraffin-embedded tissue specimens was even more hampered by fixation procedures because of the restricted epitope recognition. This was particularly evident with cross-linking fixatives such as formalin (Battifora, 1991; Battifora and Kopinski, 1986; Brandtzaeg and Rognum, 1983; Falini and Taylor, 1983; Hancock *et al.*, 1982; Larsson, 1993; Mepham, 1982). Mild fixation procedures prior to the application of monoclonal antibodies became a point of great concern, for example the optimization of formalin fixation for paraffin blocks (Battifora, 1991; De Bruijn, 1992) or the minimal fixation of cryostat sections in acetone (Warnke and Levy, 1980). However, the morphology of acetone-fixed cryostat sections is rather poor compared with formalin-fixed and paraffin-embedded tissue specimens. Therefore, a compromise between good tissue morphology and antigen preservation always has to be made (Farmilo and Stead, 1989; Hall *et al.*, 1987; Van der Loos, 1998; Van Noorden, 1986). For example, we have obtained good immunostaining results using methacarn, (see *Appendix B.1.7*) as coagulant-fixative for paraffin tissue blocks (Van der Loos, unpublished observation). A combination of standard acetone fixation followed by a 1–2 minute Zamboni fixation has been shown to be a good compromise for cryostat sections (see *Appendix C.1.1*) (Van der Loos, 1998).

The latest important event in immunohistochemistry was the introduction of heat-induced antigen retrieval. It was found that heating paraffin sections in a microwave oven, prior to the application of immunoreagent steps, resulted in a remarkable improvement in staining intensity for many antigens/epitopes, especially after using cross-linking fixatives (Shi *et al.*, 1991). Shortly after the original introduction using toxic lead salts, the buffer composition and pH, the procedure and the heating source were the subject of many studies (e.g. Cattoretti *et al.*, 1993; Evers and Uylings, 1994; Norton, 1993; Shi *et al.*, 1995; Suurmeijer and Boon, 1993). Citrate buffer (10 mM pH 6.0) came out as the most generally used heat-induced antigen retrieval reagent (Cattoretti *et al.*, 1993), although it is accepted that an optimal buffer composition and pH should be tested for individual antigens/epitopes (DAKO Guide, 1997; Shi *et al.*, 1995; Werner *et al.*, 1996). Recently,

1 mM EDTA in 10 mM Tris–HCl buffer pH 9.0 has been claimed to be more effective than citrate buffer for particular antigens/epitopes (Vyberg and Nielsen,1998; Werner *et al.*, 1996). To date, it has not been fully elucidated why the heating of tissue sections in a microwave oven, autoclave or pressure cooker improves the immunohistochemical staining properties of particular antigens (Morgan *et al.*, 1997; Shi *et al.*, 1997). Nevertheless, it is a fact that upon the application of heat-induced antigen retrieval, many antigens/epitopes which could previously not be visualized in formalin-fixed and paraffin-embedded tissue specimens, can be readily detected. When using the above-mentioned coagulant-fixative methacarn, heat-induced antigen retrieval is not beneficial or may result in a decreased immunostaining intensity (Van der Loos, unpublished observation).

1.2 Marker enzymes

To visualize antigen- or epitope-specific antibodies, a number of immunohistochemical detection methods and tracers have been proposed (*Table 1.1*). The type and number of enzymatic tracers exploited for immunoenzyme histochemistry has hardly changed since their original introduction. Horseradish peroxidase (HRP) (Nakane and Pierce, 1967) and to a lesser extent calf intestinal or *E. coli* alkaline phosphatase (AP) (Avrameas, 1969a) are still the most frequently used enzymatic markers in immunohistochemistry. Glucose oxidase from *Aspergillus niger* (Avrameas, 1969b) and *E. coli* β-galactosidase (GAL) (Bondi *et al.*, 1982) are only rarely applied. Various histochemical methods have been proposed to detect the enzyme activities, resulting in a wide variety of coloured reaction products. In this handbook the most contrasting colour combinations for immunoenzyme multiple-staining procedures with HRP, AP and GAL as tracers are shown.

1.3 Detection systems

In general, investigators searched for methods which improved the sensitivity/efficiency of the immunohistochemical staining technique: the higher the enzyme/antigen ratio, the more coloured the reaction product.

At first, the harsh reaction conditions during conjugation of enzymes to antibodies and the size of the antibody–enzyme complex after glutaraldehyde conjugation were the major concerns. Sternberger *et al.* (1970) circumvented conjugation by the introduction of the unlabelled peroxidase anti-peroxidase complex. Boorsma (1983, review) refined the conjugation procedure considerably, resulting in high-quality antibody–enzyme conjugates.

Next, in the search for more sensitivity, the extreme high-binding constant of biotin for the avidin protein was exploited (Guesdon *et al.*, 1979). Hence, methods could be adopted with much higher dilution of the primary antibody, rendering very low non-specific background signals (Coggi *et al.*, 1986; Hsu *et al.*, 1981).

The biotin–tyramide system for signal amplification originally introduced for immunoenzymatic assays (Bobrow *et al.*, 1989), appeared to also be successful in immunohistochemistry (Adams, 1992; Werner *et al.*, 1996). Until now, the rather laborious biotin–tyramide amplification, also known as catalyzed reporter deposition (CARD) has not been considered as a routinely applied detection system in immunoenzyme histochemistry. The technique appeared to be particularly useful in improving the poor staining intensities with some primary antibodies, and multi-target non-radioactive *in situ* hybridization using a sequential double immunofluorescence method (Kerstens *et al.*, 1995; Raap *et al.*, 1995).

The most recent and promising development for improving the sensitivity/efficiency of immunoenzyme histochemistry detection systems is the introduction of dextran polymer technology by DAKO A/S researchers (Bisgaard *et al.*, 1993; Bisgaard and Pluzek, 1996). Enzymatic tracers and antibodies are both coupled to a large dextran polymer molecule, ensuring a high enzyme tracer/antigen ratio. This technology is applied for primary antibodies resulting in a one-step procedure (DAKO EPOS^{TM}), as well as for secondary antibodies resulting in a two-step staining procedure (DAKO EnVision^{TM}). These new detection systems provide good staining intensity, circumvent possible problems with endogenous biotin (Sabattini *et al.*, 1998; Vyberg and Nielsen, 1998) and reduce a three-step streptavidin–biotin method to a two-step (EnVision^{TM}) or a one-step (EPOS^{TM}) user-friendly procedure.

1.4 Immunogold/silver techniques

Colloidal gold-labelled antibodies were initially developed for and applied to immunoelectron microscopy (Faulk and Taylor, 1971). As the small gold particles were not visible with light-microscopy, Holgate *et al.* (1983) adopted the silver enhancement from Danscher (1981) for the first immunogold/silver staining technique on tissue specimens. The small gold particles served as a starting point for a silver amplification of up to 40–80 nm, making the particles visible as brown/black deposits with light-microscopy. At that time however, the smallest available gold particle was 5 nm, excluding a good tissue penetration and effective silver enhancement. These problems were overcome with the development of gold particles with a size < 1 nm (Leunissen *et al.*, 1989; Van de Plas and Leunissen, 1989). In contrast with the original Danscher silver-enhancement protocol, the commercially available Aurion silver-enhancement kit is versatile and light insensitive. It can be used for the enhancement of ultra-small gold particles in both light- and electron-microscopy applications. A special feature of the silver precipitate end product, is the observation with dark field epi-polarization microscopy (De Waele *et al.*, 1988). This dark-field option is believed to be one of the most sensitive/efficient immunohistochemical detection methods available. Furthermore, it formed the basis of a double-staining method in combination with a fluorescent alkaline phosphatase detection technique (Van der Loos and Becker, 1994).

Table 1.1. Immunohistochemical light-microscopical staining techniques in their historical order

Direct/indirect peroxidase technique	Nakane and Pierce, 1967
Unlabelled antibody peroxidase technique	Avrameas, 1969b; Mason *et al.*, 1969
Indirect alkaline phosphatase technique	Avrameas, 1969a
Peroxidase anti-peroxidase (PAP) technique	Sternberger *et al.*, 1970
Alkaline phosphatase anti-alkaline phosphatase (APAAP) technique (polyclonal)	Mason and Sammons, 1978
Hapten anti-hapten techniques	Wofsy *et al.*, 1978
Labelled (strept)avidin–biotin technique	Guesdon *et al.*, 1979
Double bridge PAP technique	Vacca *et al.*, 1980
(Strept)avidin–biotin complex (SABC) technique	Hsu *et al.*, 1981
Labelled antigen method	Falini *et al.*, 1982
Immunogold/silver (IGS) technique	Holgate *et al.*, 1983
APAAP technique (monoclonal)	Cordell *et al.*, 1984
Ultra-small gold conjugates	Van de Plas and Leunissen, 1989
Biotin–tyramide amplification	Adams, 1992
Dextran polymer technology (DAKO EPOS,™ EnVision™)	Bisgaard *et al.*, 1993

1.5 Immunofluorescence

Although the clear and sharp localization of antigens with single- and double-immunofluorescence techniques (Brandtzaeg, 1998; Brandtzaeg *et al.*, 1997; Pryzwansky, 1982) is beyond dispute, these techniques still suffer from well-known drawbacks: quenching of the fluorescence signal at excitation, fading of fluorescence signal upon room temperature storage of specimens, and the occurrence of auto-fluorescence caused by formaldehyde fixation. This prevents the application of fluorescence for retrospective studies (Mason *et al.*, 1983; Mesa-Tejada, 1977; Naiem *et al.*, 1982; Taylor, 1978, 1980; Valnes and Brandtzaeg, 1982). Despite many technical improvements in this regard, the stability of a fluorescence signal has not come even close to that of permanent enzymatic reaction products. Therefore, light-microscopical immunohistochemical techniques using enzymatic chromogens often are preferred. The success of immunoenzyme histochemistry in the field of histopathology certainly did not result in extinction of fluorescence techniques. On the contrary, fluorescence techniques are still vivid and alive as reviewed by Brandtzaeg *et al.* (1997). In medical biology, a new generation of improved fluorescent labels and dyes has been introduced, mainly evolving with fluorescent-activated cell sorter techniques (FACS), confocal laser scanning microscopy techniques, multiple target *in situ* hybridization (Raap *et al.*, 1995) or combinations of *in situ* hybridization and immunohistochemistry methods (Speel *et al.*, 1995). The introduction of fluorescent enzymatic reaction products such as ELF (Larison *et al.*, 1995) by Molecular Probes Inc. (Eugene, OR), CAS Red (Becton Dickinson) and Vector Red (Vector Laboratories) (Van der Loos and Becker, 1994) has more or less unified immunoenzyme histochemistry and immunofluorescence.

1.6 Multiple immunoenzyme staining

The desire for multiple antigen visualization in one tissue specimen is almost as old as immunohistochemistry itself. In 1957 Silverstein published the first double immunofluorescence technique, shortly afterwards a second fluorochrome beside fluorescein isothiocyanate (FITC) was introduced. Almost the same has happened with immunoenzyme histochemistry. When Nakane and Pierce (1967) described their first immunoenzyme single-staining procedure, they already mentioned the possibility of multiple staining. Nakane (1968) proposed a multiple-staining method applying three indirect immunoperoxidase techniques sequentially. In this way, the localization pattern of three different antigens in one tissue section could be distinguished by three different coloured peroxidase reaction products.

In general, any immunohistochemical multiple-staining technique is a combination of multiple individual antigen detection methods. For a successful multiple-staining protocol two main problems have to be overcome. First, 'how to prevent cross-reaction(s) between both individual detection methods?', and, second, particularly for double immuno-enzymatic methods: 'which colour combinations provide the best contrast between both individual colours and a mixed-colour at sites of co-localization?' Many investigators have tried to answer these questions, resulting in a variety of double-staining procedures and colour combinations. Unfortunately, most of them are not generally applicable for a wider audience. Therefore, in this handbook only those methodologies which have the potential for 'general applicability' will be discussed.

2 Introduction to double staining

2.1 Serial sections

To complete an immunophenotypic study, there is a regular demand for the detection of more than one antigen in a single tissue specimen. The traditional application of serial sections for this purpose (Larsson, 1988; Van Noorden, 1986) can certainly be regarded as reliable but is extremely laborious. Moreover, the serial sections must be very thin (1–2 μm) to avoid the risk that tiny structures or small cell types may not be present in both sections; this requires tissue fixation and special embedding techniques possibly destroying antigens/epitopes. The 'cell-to-cell' comparison within homogeneous structured tissues, as, for example lymphoid tissue, is complicated because of the lack of tissue landmarks. Interesting co-localizations or cell-to-cell interactions will be missed easily. Cytospin, cell smears, cell culture and imprint specimens cannot be studied in this way for obvious reasons. The performance of reliable multiple staining in one tissue specimen is the best alternative to overcome these disadvantages.

2.2 Co-localization: the definition

Apart from staining two different cell types in a single tissue specimen (*Plates 1* and *7A*), the observation of co-localization (i.e. the presence of two antigens in one cell) is one of the main reasons for the performance of double staining. When the two antigens are present in the same cellular compartment, co-localization will be marked by a mixed colour, for example the co-expression of vimentin and cytokeratin in a tumour cell as shown in *Plate 2*. When the two antigens are present in different cellular compartments, co-localization will be observed as two different colours, for example nuclear positivity of a proliferation marker in a cytokeratin-positive cell (*Plates 3* and *4*). It is obvious therefore, that for proper identification of co-localization the advantages and disadvantages of the procedure of choice should be considered carefully. This handbook provides the background, necessary to perform adequate and reliable staining methods, together with practical notes that allow successful application of double-staining techniques.

2.3 Aims of double staining

- Staining two different cell types to visualize a direct overview of their localization in the tissue, for example possible cell-to-cell spatial contacts will be revealed. Double staining using primary antibodies known *a priori* to be present in two different cell types will result in two different colours, co-localization will not be present.
- Comparison of the staining patterns of a newly developed primary antibody with an antibody of a known cellular distribution. Double staining may result in either two differently coloured cell types or co-localization.
- Comparison of two primary antibodies with possible similarities. The difference or similarity of the staining patterns of two commercially available monoclonal antibodies for a particular tissue type is not always clear from the data sheets. Double staining may provide information about this feature. The presence of co-localization marked by a mixed colour is highly likely to occur.
- Comparison of immunohistochemistry and *in situ* hybridization. These two staining techniques allow, for example, the investigation of the cellular phenotype of viral-infected cells (Van den Brink *et al.*, 1990; Van der Loos *et al.*, 1989b). Combination of these two techniques in one tissue section also enables comparison of localization of a particular antigen at both mRNA and protein level (Breitschopf and Suchanek, 1996).
- Comparison of immunohistochemistry and terminal deoxytransferase (TdT)-mediated dUTP nick end labelling (TUNEL) for detection of apoptosis. Combination of these two staining techniques allows the visualization of nuclei of the cellular phenotype undergoing programmed cell death. *Chapter 7* pays special attention to this subject.
- Saving tissue sections or cell specimens

2.4 Aims of this handbook

To fulfil the aims listed above, there are three major approaches: double immunofluorescence, double immunogold techniques and double immunoenzyme techniques. This handbook is restricted to the use of multiple-staining techniques with three different enzymatic activities and a combination of an enzymatic and immunogold silver staining technique. Not included are double fluorescence techniques which are excellently reviewed by Brandtzaeg (1998), Brandtzaeg *et al.* (1997) and Speel *et al.* (1995). Also not included is multiple staining with different sizes of gold particles as applied in immunoelectron microscopy. This technique, including detailed protocols, is excellently covered in several chapters and handbooks (Geuze *et al.*, 1981; Marijianowski *et al.*, 1996; Polak and Varndell, 1984). Furthermore, double staining with two fluorochromes or gold particles of different sizes, as well as exotic combinations of immunoenzyme and immunofluorescence techniques (Ghandour *et al.*, 1979; Lechago *et al.*, 1979), will not be discussed further.

From a historical point of view, specific aims demand the creation of specially tailored double-staining methods. This included, for example, the conjugation of primary antibodies (Boorsma, 1984) or secondary antibodies (Chaubert *et al.*, 1997). Despite successful staining results, such double-staining procedures cannot be regarded as generally applicable, or reproducible by a large audience. Therefore, the choice of methods and materials in this handbook is focused on reproducible and generally applicable double-staining techniques applying commercially available reagents ('good laboratory practice'). Six double-staining methods are shown in protocol format, including suggestions for dilutions of second/third step reagents.

For the performance of immunoenzyme double staining many different colour combinations with different marker enzymes have been proposed (Van der Loos *et al.*, 1993). In this handbook attention is payed to seven colour combinations, including the different staining efficiencies of the different chromogen systems.

2.5 What are the demands for double staining?

The most important features in the successful performance of double-staining techniques are:

- two different immunohistochemical detection systems,
    (i) which do not show any cross-reactivity
    (ii) which are preferably commercially available;
- two different visualization methods
    (i) which have to show optimal colour contrast
    (ii) which allow discrimination between a mixed colour at sites of co-localization, and the two basic colours.

Every reproducible and generally applicable double-staining method, irrespective of the antibody tracers applied (haptens, enzymes, gold particles, fluorochromes) has to fulfil these two demands. The optimal conditions mentioned for the second demand are especially formulated for enzymatic chromogens.

These important demands, including the optimal conditions, are applied to the four double-staining concepts in *Chapter 3* and the seven colour combinations in *Chapter 5*.

3 Double-staining concepts

The double-staining concepts described here have been proven to fulfil the major demands mentioned in *Section 2.5*, and can be performed using commercially available reagents. Detailed 'blueprints' of these five concepts including suggestions for dilution of second/third-step reagents are found in *Appendix D.1.1–1.5*. The figures between brackets in the schematic drawing of the double-staining concepts correspond to the subsequent incubation steps in the protocols in Appendix D.

3.1 Sequential techniques

Sequential double-staining techniques involve two complete immuno-enzymatic staining procedures performed one-by-one. The main problem with this concept for double staining is preventing cross-reaction of immunoreagents between the first and second staining sequence. Numerous modifications of *sequential* staining techniques with or without removal of the first set of reagents have been described.

- Antibody elution step after the completion of the first staining sequence. For this purpose either acidic solutions (Nakane, 1968), oxidation with a permanganate solution (Tramu *et al.*, 1978), chaotropic reagents, such as 1 M ammoniumisothiocyanate (Van der Loos, unpublished observation), or electrophoresis (Vandesande, 1983) have been proposed. However, when applying acidic solutions, high-affinity primary antibodies may remain at their binding site (Tramu *et al.*, 1978, Vandesande, 1983).
- The use of 3,3′-diaminobenzidin (DAB) as chromogen for HRP activity in the first staining sequence results in an effectively sheltering reaction product. When the first set of immunoreagents is completely covered with DAB-polymer, the second step reagents are unable to cross-react (Sternberger and Joseph, 1979). It is claimed that complete sheltering will also be provided by the immunogold/silver end-product (Krenács *et al.*, 1990). It cannot be excluded, however, that effective sheltering by the DAB end-product may also hide the second antigen, especially if both antigens are in close proximity to

each other (Valnes and Brandtzaeg, 1984). Other HRP chromogens such as 3-amino-9-ethylcarbazole (AEC), tetramethylbenzidine (TMB) and alkaline phosphatase (AP), β-galactosidase (GAL) chromogens do not have this sheltering characteristic, or at least have it to a lower extent.

- The large polymeric structure of StrepABComplex/HRP or DAKO EPOS^{TM} reagents contributes to the sheltering effect by steric hindrance (Krenács *et al.*, 1990; Mullink *et al.*, 1987; Pastore *et al.*, 1995).
- The first set of reagents can be removed very effectively by a heat-induced antigen retrieval procedure with citrate (100 mM, pH 6.0) (Brandtzaeg *et al.*, 1997; Lan *et al.*, 1995). One should be aware that not all reaction products are capable of surviving this boiling step. For example, the staining accuracy of DAB, AEC, Fast Red, 5-bromo-4-chloro-3-indolyl-β-D galactoside (X-GAL) chromogens remained unchanged after boiling, whereas nitro blue tetrazolium (NBT)/5-bromo-4-chloro-3-indolyl-phosphate (BCIP) reaction product became extremely blurry (Van der Loos, unpublished observation). The boiling method for removing the first set of reagents is obviously only applicable for paraffin sections.

Sequential double-staining techniques are useful for the evaluation of two different cell populations or cell constituents. The recently available DAKO EnVision^{TM} Doublestain System is based on polymer technology (Heras *et al.*, 1995) and provides a user-friendly system for *sequential* double staining. *Figure 3.1* represents a schematic drawing of this commercially available kit system corresponding to the protocol in *Appendix D.1.1*. A typical example using two mouse monoclonal antibodies is shown in *Plate 4*.

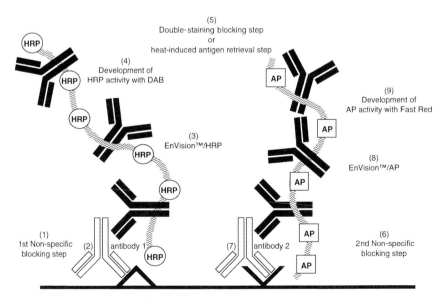

Figure 3.1. DAKO EnVision^{TM} Doublestain System (sequential concept).

Important: one should be aware that *sequential* double-staining techniques always inherit the risk of spurious double-stained structures. Therefore, these techniques are not recommended for those instances where mixed-coloured products are expected at sites of co-localization.

3.2 Simultaneous technique: *direct/direct*

In a *direct/direct* concept both primary antibodies are conjugated differently. The tracers directly linked to primary antibodies can be either enzymes, biotin (with a streptavidin reagent as second layer), haptens (with anti-hapten as second layer) or fluorochromes (Behringer *et al.*, 1991; Boorsma, 1984; Falini *et al.*, 1986; Mason *et al.*, 1983; Newman and Jasani, 1984; Van der Loos *et al.*, 1987, 1989a; Wallace and Wofsy, 1979). When using fluorochrome-conjugated antibodies for immunoenzyme double staining, the fluorochrome label is no longer applied for light emission, but as a hapten (with anti-fluorochrome antibodies). The *direct/direct* double-staining concept is completely independent of the primary antibody species, Ig isotype or IgG subclass. In contrast with the *sequential* double-staining concepts, time-saving antibody cocktails can be applied for simultaneous double-staining methods. Whenever one particular primary antibody needs a 60 min room-temperature incubation and the other primary antibody incubation overnight at 4°C, sequential incubations are still possible; the following secondary antibodies can again be incubated in a cocktail. The two enzymatic activities are developed sequentially, after finishing all antibody and detection steps. *Figure 3.2* is a schematic drawing of the *direct/direct* concept, corresponding with the protocol in *Appendix D.1.2*,

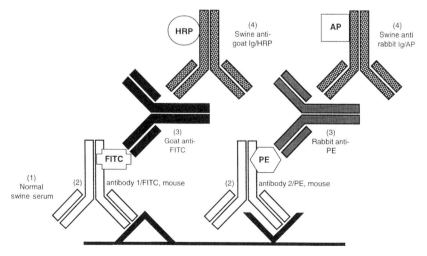

Figure 3.2. *Direct/direct* simultaneous double-staining concept.

applied for the double-staining combination with two fluorochrome-labelled primary antibodies, and visualized with HRP and AP as tracer enzymes (*Plate 5*).

3.3 Simultaneous technique: *indirect/indirect*

Indirect/indirect double-staining concepts are performed with two primary antibodies raised in different species (Campbell and Bhatnagar, 1976; Mason and Sammons, 1978). Again as with *direct/direct* double-staining concepts time-saving antibody cocktails can be applied, and the two enzymatic activities are developed last. It is recommended that secondary antibodies raised in the same host are applied, in order to prevent unexpected interspecies cross-reactions. Whenever the regular 2-step detection procedures provide too low sensitivity/efficiency (*Figure 3.3*), a 3-step procedure with a biotinylated second step and enzyme-conjugated streptavidin can be applied. Alternatively, a DAKO En-Vision™ reagent can be used for enhancing a 2-step peroxidase technique (*Figure 3.4*).

Plates *1, 3, 6* and *10* are examples of the *indirect/indirect* double-staining concept with two primary antibodies of different animal species. Schematic drawings of this procedure are given in *Figures 3.3* and *3.4* corresponding with the basic protocol in *Appendix D.1.3*.

An elegant variant on this *indirect/indirect* double-staining concept can be performed with two monoclonal antibodies from different Ig isotypes or IgG subclasses (Tidman *et al.*, 1981). Commercially available secondary reagents directed against mouse Ig subtypes IgG and IgM, or IgG isotypes

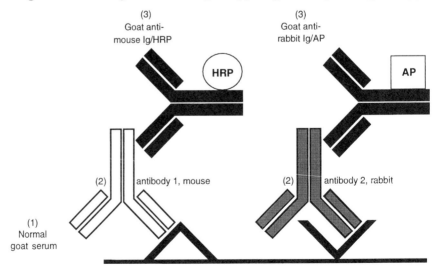

Figure 3.3. *Indirect/indirect* simultaneous double-staining concept.

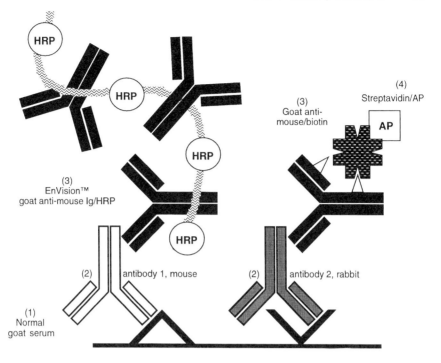

Figure 3.4. *Indirect/indirect* simultaneous double-staining concept, with enhancements.

1, 2a, 2b and 3 have been shown to react specifically (Southern Biotechnology Associates Inc.). Be aware of the fact that staining sensitivity/efficiency with anti-IgG subclass-specific antibodies is somewhat lower compared with antibodies directed against the whole Ig-molecule as normally used for single staining. This lower efficiency can be compensated by using either a higher concentration of primary antibody (2–4 times), and/or a more sensitive detection method (*Table 5.2*), and/or using an overnight incubation at 4°C instead of 60 min at room temperature.

Plate 7A shows an example of the *indirect/indirect* concept with two mouse primary antibodies of respectively IgG3 and IgG2a subclass. A schematic drawing of that procedure is provided in *Figure 3.5*, corresponding with the basic protocol in *Appendix D.1.4*.

3.4 Multi-step technique: *indirect/direct*

The *indirect/direct* double-staining concept is based on the application of an unlabelled primary antibody in combination with a conjugated primary antibody of the same species and Ig isotype or IgG subclass. This second primary antibody can be either enzyme-conjugated (Van der

(Chapter 3 continues on p. 33)

Plate 1. Simultaneous *indirect/indirect* double-staining method (*Appendix D.1.3*) performed with mouse anti-CD79a, JCB117 marking B lymphocytes/plasma cells, and rabbit anti-CD3 marking T lymphocytes as primary antibodies (both DAKO). Formalin-fixed paraffin section from an abdominal aorta segment with aneurysm, showing detail of an adventitial lymphocyte infiltrate. The section was pretreated with heat-induced antigen retrieval using citrate pH 6.0. CD79a in blue (alkaline phosphatase, Fast Blue) and CD3 in brown (peroxidase, DAB). Bar = 50 μm.

Plate 2. Multi-step *indirect/direct* double-staining method (*Appendix D.1.5*) performed with mouse anti-vimentin, V9 and FITC-conjugated mouse anti-cytokeratin, MNF116 as primary antibodies (both DAKO). Acetone-fixed cryostat section from kidney adenocarcinoma (Grawiz tumour) showing a tumour cell co-expressing both cytoskeleton markers stained in a purple mixed colour. Vimentin in blue (alkaline phosphatase, Fast Blue) and cytokeratin in red (peroxidase, AEC). Bar = 50 μm.

Plate 3. Simultaneous *indirect/indirect* double-staining method (*Appendix D.1.3*) performed with mouse anti-cytokeratin, MNF116 and rabbit anti-Ki-67 as primary antibodies (both DAKO). Formalin-fixed paraffin section from a breast carcinoma showing proliferating nuclei, pretreated with heat-induced antigen retrieval using citrate pH 6.0. Cytokeratin in blue (alkaline phosphatase, Fast Blue) and Ki-67 in red (peroxidase, AEC). Bar = 50 μm.

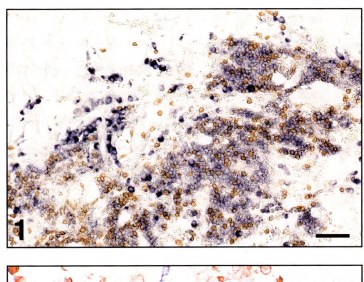

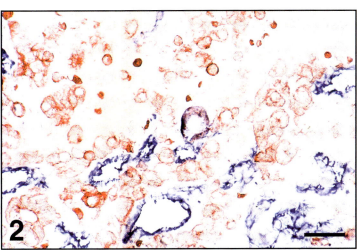

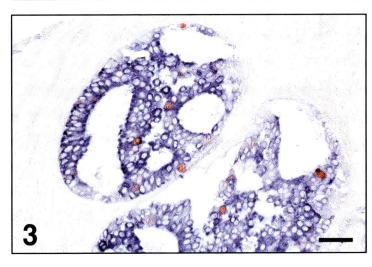

Plate 4. *Sequential* double-staining method (*Appendix D.1.1*) illustrating the DAKO EnVision™ Doublestain System using mouse anti-p53, DO-7 and mouse anti-cytokeratin, AE1/AE3 as primary antibodies (both DAKO). Formalin-fixed paraffin section from a colon carcinoma, pretreated with heat-induced antigen retrieval using citrate pH 6.0. Cytokeratin in brown (peroxidase, DAB) and p53 in red (alkaline phosphatase, DAKO Fast Red). Nuclear counterstain with haematoxylin. Bar = 50 μm.

Plate 5. Simultaneous *direct/direct* double-staining method (*Appendix D.1.2*) performed with Phycoerythrin-conjugated mouse anti-CD3, SK7 (Leu 4) marking T lymphocytes and FITC-conjugated mouse anti-CD25, 2A3 (IL-2 receptor) marking Interleukin-2 Receptor as primary antibodies (both Becton Dickinson). Acetone + Zamboni-fixed cryostat section from a normal tonsil of a 4-year-old child. T cells in blue (alkaline phosphatase, Fast Blue) and IL-2R in red (peroxidase, AEC). Activated T lymphocytes show a purple co-localization. Bar = 50 μm.

Plate 6. Simultaneous *indirect/indirect* double-staining method (*Appendix D.1.3*) performed with mouse anti-CD34, QBEnd10 marking endothelial cells and rabbit anti-von Willebrand factor (vWf) also marking endothelial cells as primary antibodies (both DAKO). Acetone + Zamboni-fixed cryostat section from a tonsil of a young child. CD34 in purple/blue (alkaline phosphatase, BCIP/NBT) and vWf in brown (peroxidase, DAB). Co-localization is regularly seen, but CD34 is also found at vessels and capillaries which are negative for vWf. Nuclear counterstain with Methylgreen; the section is mounted organically. Bar = 50 μm.

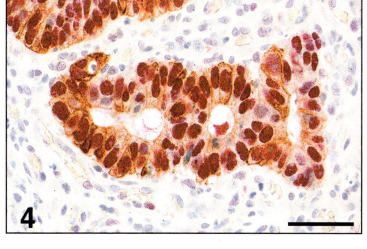

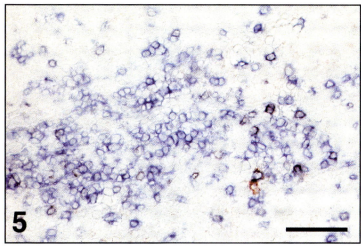

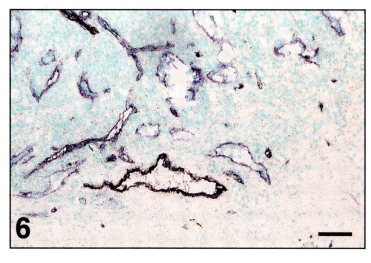

Plate 7. Simultaneous *indirect/indirect* double-staining method (*Appendix D.1.4*) performed with mouse anti-CD68, PG-M1 (IgG$_3$) marking macrophages, and mouse anti-α-actin, 1A4 (IgG2$_a$) marking smooth muscle cells (both DAKO). Formalin-fixed paraffin section from a carotis segment, showing a detail of an atherosclerotic lesion composed of smooth muscle cells in turquoise (β-galactosidase, X-GAL) and macrophages in red (alkaline phosphatase, New Fuchsin). Section was pretreated with heat-induced antigen retrieval using citrate pH 6.0. Bar = 100 μm.

A–C: Bright field microscopy, original picture (A), with 622 nm narrow band pass filter (B), with 478 nm narrow band pass filter (C), respectively. Note the unwanted imaging of blue cells in the upper half of *7B*.

D–F: Image analysis of the same double staining specimen with Adobe Photoshop 3.0 according to Lehr *et al.* (1999). D, Colour contrast in *7A* is enhanced using the 'hue/saturation' tool from the 'image/adjust' menu. Selecting the total image gives the total number of pixels read in the 'histogram' command from the 'image' menu. E–F: The specific colours can be selected individually using the 'magic wand' tool and the commands 'grow' and 'similar' in the 'select' menu. Pixel number can be read again after the 'histogram' command. In this example the pixel counts for the turquoise and red chromogens were 25 425 and 158 125, representing respectively 4.6 and 28.3% of the surface of the total image in *7D*.

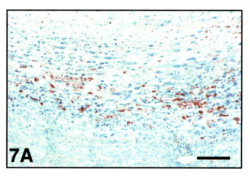

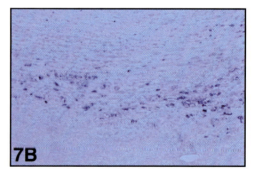

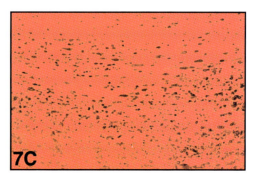

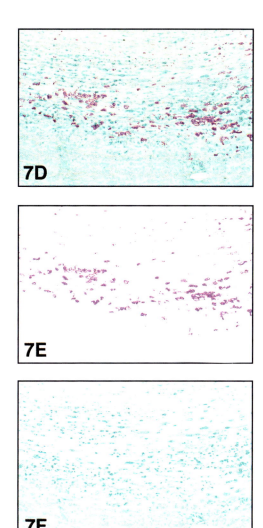

Plate 8. Multi-step *indirect/direct* double-staining method (*Appendix D.1.5*) performed with mouse anti-CD4, MT310 marking T helper/inducer lymphocytes and FITC-conjugated mouse anti-CD8, DK25 marking T suppressor/cytotoxic lymphocytes as primary antibodies (both DAKO). Acetone + Zamboni-fixed cryostat sections from the normal thymus of a one year old child. Immature thymocytes in the cortex (C) co-express both CD4 and CD8 at the cell surface; matured thymocytes in the medulla (M) express either CD4 or CD8. Bar = 50 μm.
A: CD4 in red (alkaline phosphatase, DAKO Fast Red) and CD8 in brown (peroxidase, DAB).
B: CD8 in red (alkaline phosphatase, DAKO Fast Red) and CD4 in brown (peroxidase, DAB). Note the poor colour contrast of the mixed colour for both *8A* and *B*.
C: CD4 in purple/blue (alkaline phosphatase, NBT/BCIP) and CD8 in brown (peroxidase, DAB).
D: CD8 in purple/blue (alkaline phosphatase, NBT/BCIP) and CD4 in brown (peroxidase, DAB). Note the greyish mixed colour in the cortex in *8C/D* marking co-localization. Also note that colour contrast in *8D* is better than in *8C*, because in the latter the purple/blue stained CD4 majority population in the medulla is overwhelming the CD8 minority population in yellow/brown.
E: CD4 in blue (alkaline phosphatase, Fast Blue) and CD8 in brown (peroxidase, DAB).
F: CD8 in blue (alkaline phosphatase, Fast Blue) and CD4 in brown (peroxidase, DAB). Note the greyish mixed colour in the cortex in *8E/F* marking co-localization. This colour contrast is certainly less than *8C/D* and *8G/H*.
G: CD4 in blue (alkaline phosphatase, Fast Blue) and CD8 in red (peroxidase, AEC).
H: CD8 in blue (alkaline phosphatase, Fast Blue) and CD4 in red (peroxidase, AEC). Note the purple mixed colour in the cortex in *8G/H* marking co-localization. Note that the contrast in *8G* is better than in *8H*, because in the latter the red stained CD4 majority population in the medulla is overwhelming the CD8 minority population in blue.
I: CD4 in turquoise (β-galactosidase, X-GAL) and CD8 in red (alkaline phosphatase, Fast Red).
J: CD8 in turquoise (β-galactosidase, X-GAL) and CD4 in red (alkaline phosphatase, Fast Red). Note the purple/blue mixed colour in the cortex in *8I/J* marking co-localization. Also note that in spite of following the recommendation to stain the CD4 population in turquoise (see Section 6.1) the colour combination in *8J* shows better contrast between mixed colour and basic colours than in *8I*.
K–M: Double staining using the same pair of primary antibodies, but applying a combination of fluorescent AP and IGS staining (*Appendix D.1.6*). K: Double exposure of CD4 in red (alkaline phosphatase, Vector Red) and CD8 in blue (immunogold/silver). Note the yellow/white mixed colour marking co-localization in *8K*. L and M are individual images of the upper half of *8K*.

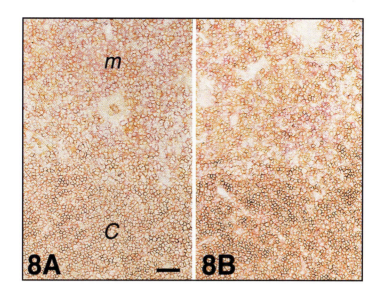

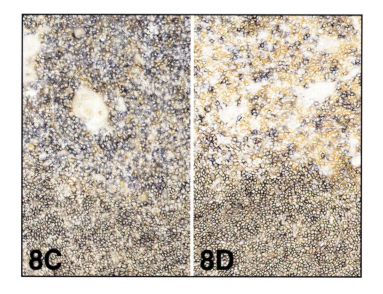

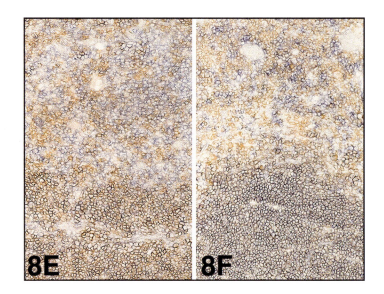

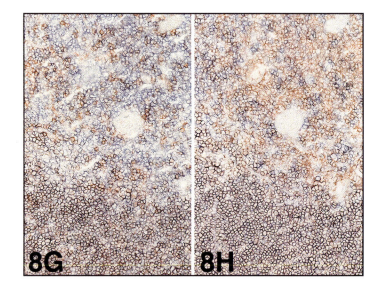

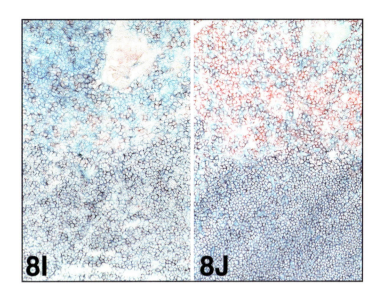

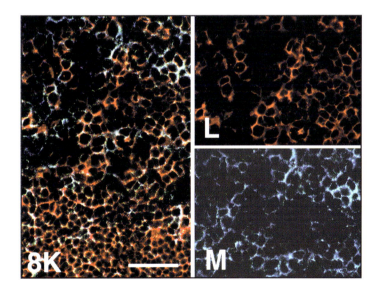

Plate 9. Multi-step *indirect/direct* double-staining method (*Appendix D.1.5*) performed with mouse anti-HLA-DR, CR3/43 marking the major histocompatibility complex class II (DAKO) and FITC-conjugated mouse anti-CD3, SK7 (Leu 4) marking T lymphocytes (Becton Dickinson) as primary antibodies. Acetone + Zamboni-fixed cryostat sections from a coronary artery. A and B are semi serial sections processed with identical primary antibodies but in reversed colour combinations. Note that the lymphocytes are much better observed in B than in A. Bar = 50 μm.
A: CD3 in blue (alkaline phosphatase, Fast Blue) and HLA-DR in red (peroxidase, AEC).
B: HLA-DR in blue (alkaline phosphatase, Fast Blue) and CD3 in red (peroxidase, AEC).

Plate 10. Simultaneous *indirect/indirect* double-staining method (*Appendices D.1.3* and *D.1.6*) performed with mouse anti-α-actin, 1A4 marking smooth muscle cells (DAKO) and rabbit anti-interleukin-1α (Genzyme/R&D Systems) as primary antibodies. Acetone + Zamboni-fixed cryostat section from an abdominal aorta segment with atherosclerosis. Smooth muscle cells in red (alkaline phosphatase, Vector Red) and interleukin-1α in blue (immunogold/silver). Double exposure of the individual images of the fluorescent alkaline phosphatase reaction product and immunogold/silver product, respectively, is merging to yellow/white at sites of co-localization as seen in the proliferating smooth muscle cells. Bar = 25 μm.

Plate 11. Simultaneous visualization of mouse anti-CD20, L26 marking B-lymphocytes and TUNEL performed according to the protocol in *Appendix D.1.7*. After TUNEL there was no tissue pretreatment procedure. Formalin-fixed paraffin tissue section from tonsil, showing a detail of CD20 positive B cells in brown (peroxidase, DAB) in a lymphoid follicle centre with adjacent B-cell zone, and apoptotic bodies in blue/purple (alkaline phosphatase, NBT/BCIP). Apoptosis is mainly found in the follicle centre. Nuclear counterstain with Methylgreen. Bar = 100 μm.

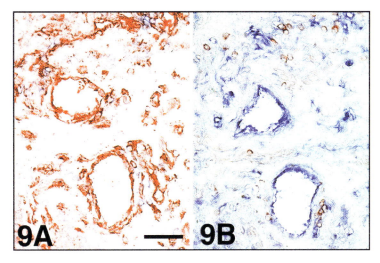

9A **9B**

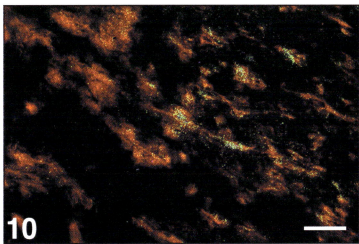

10

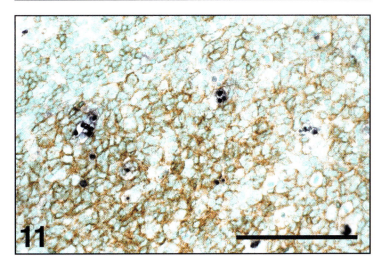

11

Plate 12. Simultaneous visualization of mouse anti-CD68, PG-M1 marking macrophages and TUNEL performed according to the protocol in *Appendix D.1.7.* After TUNEL the section was pretreated with heat-induced antigen retrieval using citrate pH 6.0. Formalin-fixed paraffin tissue section from tonsil, showing a detail of a lymphoid follicle centre with CD68-positive starry sky macrophages in brown (peroxidase, DAB), and apoptotic bodies in red (alkaline phosphatase, Fast Red). TUNEL-positivity is clearly shown inside the macrophages, which have scavenged the apoptotic B cells. The arrowhead indicates a starry sky macrophage which is shown at high magnification in the insert. Nuclear counterstain with haematoxylin. Bar = 100 μm.

Plate 13. Triple-staining technique according to the protocol in *Appendix H.1.3* consisting of a simultaneous *indirect/indirect/indirect* method based on different mouse IgG subclasses combined with a rabbit antibody. Mouse anti-α-actin, 1A4 (IgG$_{2a}$) marking smooth muscle cells (DAKO), mouse anti-collagen type III, HWD1.1 (IgG$_1$) marking connective tissue (Biogenex), and rabbit anti-Von Willebrand factor marking endothelial cells (DAKO) were used as primary reagents. Acetone + Zamboni-fixed cryostat section, showing a detail of the myocardium including a side branch of a coronary artery. Smooth muscle cells in turquoise (β-galactosidase, X-GAL), connective tissue in blue (alkaline phosphatase, Fast Blue) and endothelial cells in red (peroxidase, AEC). Bar = 50 μm.

Plate 14. Triple-staining technique according to the protocol in *Appendix H.1.3* consisting of a simultaneous *indirect/indirect* method based on different mouse IgG subclasses combined with a multi-step *indirect/direct* method. Mouse anti-α-actin, 1A4 (IgG$_{2a}$) marking smooth muscle cells (DAKO), mouse anti-CD68, EBM11 (IgG$_1$) marking macrophages (DAKO), and FITC-conjugated mouse anti-CD3, SK7 (IgG$_1$) marking T lymphocytes (Becton Dickinson) were used as primary reagents. Acetone + Zamboni-fixed cryostat section from an aorta segment, showing a detail of an atherosclerotic lesion. Smooth muscle cells in turquoise (β-galactosidase, X-GAL), macrophages in blue (alkaline phosphatase, Fast Blue) and T lymphocytes in red (peroxidase, AEC). Bar = 50 μm.

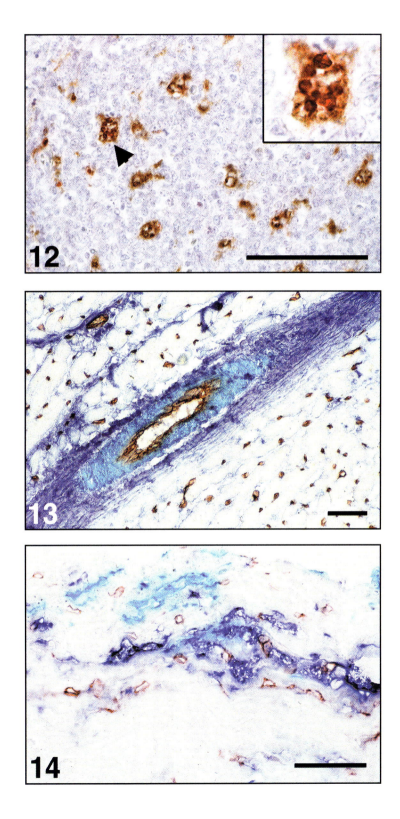

Plate 15. Triple-staining technique according to the protocol in *Appendix H.1.3* consisting of a *sequential* double-staining method combined with the *indirect/indirect* method based on different animal species. Rabbit anti-secretory mucin 6 [De Bolos *et al.* (1995) *Gastroenterology* **109**: 723], mouse anti-secretory mucin 5AC (Novocastra Ltd.), and rabbit anti-*Helicobacter pylori* (DAKO) were used as primary antibodies. Anti-MUC6 antibody was a kind gift from the author. Formalin-fixed paraffin section from stomach antrum with *H. pylori* infection. In the first staining sequence MUC6 is stained in brown (peroxidase, DAB), followed by heat induced antigen retrieval. Next, MUC5AC is stained in turquoise (β-galactosidase, X-GAL), and *H. pylori* in red (alkaline phosphatase, DAKO Fast Red). *15B* is a higher magnification of the boxed area in *15A*. Note the preference of *H. pylori* for MUC5AC positive crypts. Nuclear counterstain with haematoxylin. Bar = 100 μm. Courtesy of Dr Kristien Tytgat, Dept. of Gastroenterology, Academical Medical Centre, Amsterdam and Dr Gijs van den Brink, Pediatric Gastroenterology and Nutrician Research Group, Sophia Children's Hospital, Rotterdam, The Netherlands.

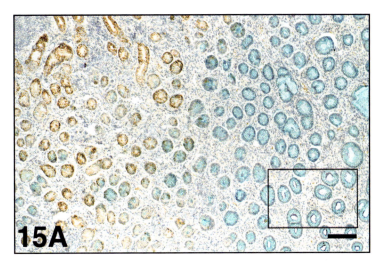

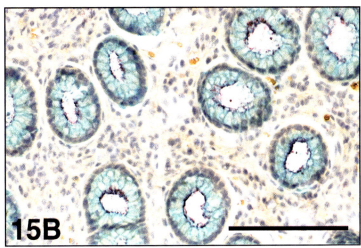

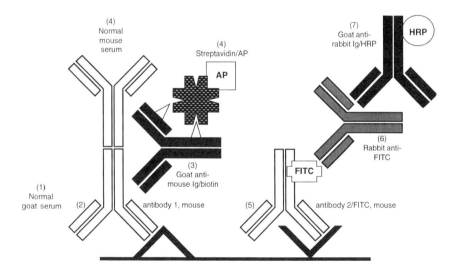

Figure 3.5. *Indirect/indirect* simultaneous double staining concept with two mouse monoclonal antibodies of different IgG subclass.

Figure 3.6. *Indirect/direct* multi-step double staining concept using streptavidin-biotin detection for the unlabelled antibody.

Loos *et al.*, 1987), biotinylated (Van der Loos *et al.*, 1988), FITC conjugated (Van der Loos *et al.*, 1989a) (*Figures 3.6–3.8*), or a DAKO EPOS[TM]/HRP reagent (Van der Loos *et al.*, 1996) (*Figure 3.9*).

The staining procedure starts with the unlabelled primary antibody, followed by appropriate detection steps, but without developing the enzymatic activity. Three different alternative detection systems for the first part of the *indirect/direct* double-staining procedure are shown in

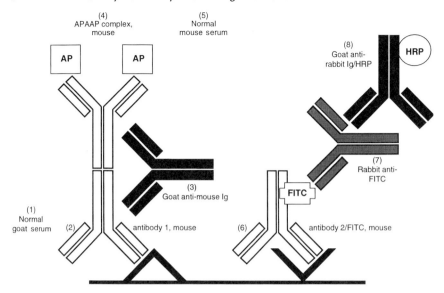

Figure 3.7. *Indirect/direct* multi-step double-staining concept using APAAP detection for the unlabelled antibody.

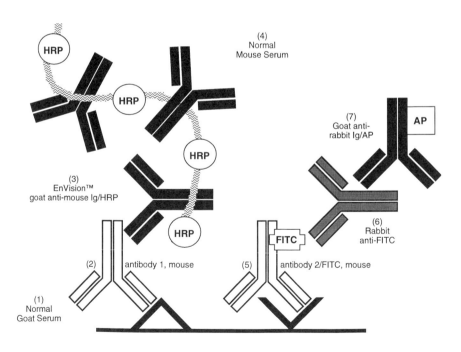

Figure 3.8. *Indirect/direct* multi-step double-staining concept using DAKO EnVision™ for the detection of the unlabelled antibody.

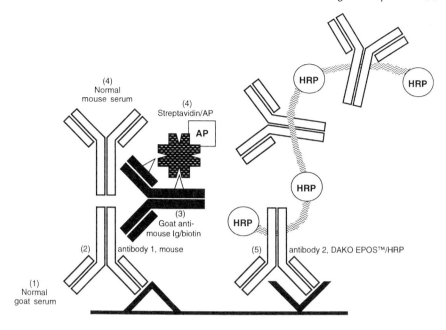

Figure 3.9. *Indirect/direct* multi-step double-staining concept using DAKO EPOS™.

Figures 3.6–3.8 and the protocols in *Appendix D.1.5.* Before the incubation with the second conjugated primary antibody, an incubation step with normal serum equal to the host of the primary antibodies is inserted. This blocking step will saturate the secondary reagent, and prevents the conjugated primary antibody from cross-reacting with the detection reagents involved with the first unlabelled primary antibody (Van der Loos *et al.*, 1987). After finishing all antibody and detection steps the two enzymatic activities are developed sequentially. The *indirect/direct* double staining concept shows high flexibility with respect to the applied enzyme- or gold-particle-conjugated reagents. This is illustrated with CD4/CD8 double staining using one primary antibody combination for 11 different colour combinations (*Plate 8A–K*).

Technical developments similar to this *indirect/direct* concept for double staining are also described by others (Gillitzer *et al.*, 1990; Glezerov, 1986; Kupper and Storz, 1986; Newman *et al.*, 1989; Oliver, 1990).

A schematic drawing of the *indirect/direct* concept is given in *Figures 3.6–3.9* and corresponds with the basic protocol in *Appendix D.1.5.*

4 Multiple-staining strategy and controls

4.1 General strategy

An immunophenotypic study usually begins by selecting a number of antibody markers relevant to the question. Occasionally, pilot experiments have to be carried out to find the optimal fixation and embedding procedure. In case of retrospective studies involving archival tissue specimens, which are fixed and embedded, the applicability of the selected antibodies should be checked (Farmilo and Stead, 1989) and/or tissue pretreatment procedures should be tested (DAKO Guide, 1997; Shi *et al.*, 1997). If multiple staining is to be a part of an immunophenotypic study, a strategy such as that shown below can be followed for reasons of efficiency.

4.1.1 Single staining

Prepare enough tissue sections from each tissue block for all immunophenotyping experiments, including single-, double- and triple staining. If possible, cut (semi-) serial sections and number them. Paraffin sections can be stored as ribbons, while cryostat sections, cell preparations, etc. can be air-dried, and stored unfixed at −80°C (*Appendix C.1.1*). After microscopical observation of all single immunostained slides, a small number of 'investigation representative' cases are selected for double-staining experiments.

4.1.2 Double staining

Sections in storage are used to perform the double staining experiments. A pilot double-staining experiment is performed first, using only a limited number of sections and including all controls (*Section 4.2*). This first experiment allows us to evaluate whether comparable staining patterns are obtained as with the original single staining experiments. Moreover, in case co-localization of two antigens or epitopes is expected, mixed-stained cell structures should be distinguished from both basic colours.

4.1.3 Triple staining

Only a few combinations of two double-staining experiments with one common antibody marker should be selected for triple staining. In our experience, triple-staining results will give a superb overview and are helpful in verifying what has already been evaluated using double-staining procedures. Suggestions for triple-staining procedures and colour combinations are discussed in *Chapter 8*.

4.2 Double-staining control experiments

Control experiments are an absolute necessity for any immunohisto-chemical investigation (Polak and Van Noorden, 1997). For example, the use of species-, Ig-isotype, or IgG-subclass irrelevant immune reagents applied in the same Ig concentration as the original primary antibody, serves as an important negative control experiment.

Apart from essential negative control and other experiments and/or tissue-related positive and negative controls, additional double-staining controls should be included in the first series of experiments. These controls consist of two specimens which undergo all incubation steps and detection procedures, but in which primary antibody 1 or 2, respectively, have been omitted (De Boer *et al.*, 1997; Mason and Woolston, 1982; Van der Loos *et al.*, 1996). These specimens should be run in parallel with the true double-staining experiment. Although other marker enzymes and chromogens may be used for double staining as with the original single-staining specimens, both halves of the double-staining protocol should each reveal a similar staining pattern and staining intensity. Moreover, these extra controls may provide additional information regarding interspecies cross-reactivity and false-positivity due to endogenous enzyme activities or unwanted binding of reagents.

On some occasions there might be doubt as to whether cellular structures are truly double stained. This is caused mainly by the fact that both markers will never be expressed at the same level in each cell. This will cause a variability in appearance of the intermediate colour at sites of co-localization. In such cases it is recommended that experiments with the reversed colour combination, or an entirely different colour combination are also performed. This may give additional information on this sometimes difficult issue. This additional reversed experiment is definitely needed when double-stained cells are the subject of an enumeration study. Counting of double-stained cells should match for both colour combinations.

5 Selecting a concept, detection system and colour combination for double staining

5.1 Selection of a double-staining concept

The aim of a double-staining method is usually formulated during the evaluation of individual single-staining specimens. This implies that the choice of an appropriate double-staining concept should be fully tailored to the characteristics of those primary antibodies applied to the original individual single-staining experiments. A flow chart for the selection of the most appropriate double-staining concept for a given combination of primary antibodies (*Table 5.1*) is based on the idea that the choice of double-staining method should be as simple as possible. For example, commercially available conjugates have preference over self-conjugated antibodies.

When obtaining a 'NO' after the last question, the most frequently encountered situation is a pair of two mouse monoclonal antibodies of subclass IgG1, both unavailable in a conjugated format (fluorochrome, biotin, or enzyme).

In such a situation there are the following alternatives:

1. Search for a replacement primary antibody either raised in a different species, or a monoclonal antibody of different isotype or IgG subclass, or perhaps an antibody which is conjugated directly with either an enzyme, hapten or fluorochrome. Always check the replacement primary antibody for its staining pattern compared with the original antibody, preferably with a double-staining experiment.
2. Although the *sequential* double-staining technique is normally not recommended for double-staining experiments that have mixed colours at sites of co-localization, it could be advantageous to test it. It has to be stressed that when applying a *sequential* double-

Table 5.1. Flow chart for selection of the simplest double-staining concept

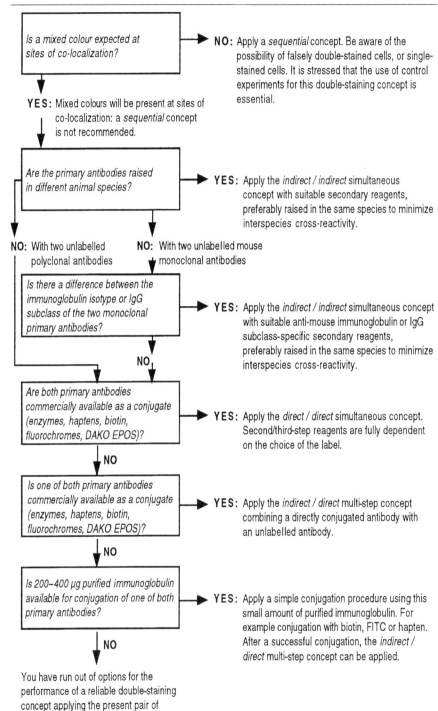

Is a mixed colour expected at sites of co-localization?

NO: Apply a *sequential* concept. Be aware of the possibility of falsely double-stained cells, or single-stained cells. It is stressed that the use of control experiments for this double-staining concept is essential.

YES: Mixed colours will be present at sites of co-localization: a *sequential* concept is not recommended.

Are the primary antibodies raised in different animal species?

YES: Apply the *indirect / indirect* simultaneous concept with suitable secondary reagents, preferably raised in the same species to minimize interspecies cross-reactivity.

NO: With two unlabelled polyclonal antibodies

NO: With two unlabelled mouse monoclonal antibodies

Is there a difference between the immunoglobulin isotype or IgG subclass of the two monoclonal primary antibodies?

YES: Apply the *indirect / indirect* simultaneous concept with suitable anti-mouse immunoglobulin or IgG subclass-specific secondary reagents, preferably raised in the same species to minimize interspecies cross-reactivity.

NO

Are both primary antibodies commercially available as a conjugate (enzymes, haptens, biotin, fluorochromes, DAKO EPOS)?

YES: Apply the *direct / direct* simultaneous concept. Second/third-step reagents are fully dependent on the choice of the label.

NO

Is one of both primary antibodies commercially available as a conjugate (enzymes, haptens, biotin, fluorochromes, DAKO EPOS)?

YES: Apply the *indirect / direct* multi-step concept combining a directly conjugated antibody with an unlabelled antibody.

NO

Is 200–400 µg purified immunoglobulin available for conjugation of one of both primary antibodies?

YES: Apply a simple conjugation procedure using this small amount of purified immunoglobulin. For example conjugation with biotin, FITC or hapten. After a successful conjugation, the *indirect / direct* multi-step concept can be applied.

NO

You have run out of options for the performance of a reliable double-staining concept applying the present pair of primary antibodies.

staining technique, false colour mixing may occur or single-stained structures where double-staining should occur (*Section 3.1*). A modern approach to the *sequential* double-staining technique has become available with DAKO EnVision™ technology (*Appendix D.1.1*).

3. Very recently, DAKO launched the A(nimal) R(esearch) K(it). This ARK kit is basically intended for the application of mouse monoclonal antibodies on mouse or rat tissue specimens, preventing the problem of cross-reacting endogeneous immunoglobulin. The principle is an *in vitro* biotinylation of the mouse primary antibody with a goat anti-mouse Fab Ig/biotin reagent, followed by blocking with normal mouse Ig. Although our experience with this ARK kit using it for double staining on human tissue is very limited, it is possible when dealing with two mouse IgG1 unlabelled primary antibodies, to biotinylate one primary antibody using the ARK kit. After biotinylation this reagent can be used in a multistep double staining protocol. So far, good results have been obtained with an unlabelled primary antibody detected with Envision™/goat anti-mouse Ig-HRP, and an ARK kit biotinylated second primary antibody which is finally detected with streptavidin/AP (see *Appendix D.1.5*). Because the DAKO ARK kit was launched at the time this handbook went to press, it should be stressed that this procedure has only been briefly tested for a limited number of antibody combinations (Van der Loos and Göbel, unpublished observations). This simple ARK kit labelling procedure replaces the far more laborious conjugation methods as mentioned in number 4.

4. Purify the immunoglobulin fraction from a large amount of one of the primary antibodies, and conjugate with FITC, biotin or a hapten (*Appendix G.1.1* and *G.1.2*). From our practical experience it appears that the 200–400 µg of immunoglobulin needed for conjugation, should be purified (Protein A method) from 4–6 vials of commercial antibody containing 100 µg each. Because this is rather expensive it should be stressed that we have encountered several primary antibodies which could not resist the mild biotinylation procedure (Van der Loos, unpublished observation). Therefore this method should be considered as optional because of its costs and reliability regarding the final conjugate.

5.2 Selecting two individual detection systems

After the selection of two non-cross-reacting immunoenzymatic methods as a general concept for double staining, a definitive selection of both detection systems depends on some additional considerations.

5.2.1 The sensitivity of both immunohistochemical techniques should be matched to the antigen density

In case of low expressing antigens/epitopes highly sensitive/effective detection systems are needed for appropriate visualization. The final staining sensitivity/efficiency consists of two components:

1. The detection system itself: current detection systems differ in their sensitivity/efficiency (DeLellis and Kwan, 1988); for example, a traditional two-step procedure is less sensitive than a three-step detection procedure (PAP, APAAP, double-bridge APAAP, Strep ABComplex methods) or a modern two-step procedure with DAKO EnVision^{TM}. Inefficient detection methods may result in negative or weak staining intensity (Bosman *et al.*, 1983; Coggi *et al.*, 1986; Shi *et al.*, 1988; Valnes *et al.*, 1984). *Table 5.2* provides an overview of current detection methods in relation to their sensitivity/efficiency according to our insights.

2. Some substrate/chromogen systems will produce a reaction product more efficiently than others (De Jong *et al.*, 1985; Scopsi and Larsson, 1986; Trojanowski *et al.*, 1983). *Table 5.3* provides an overview of current enzymatic visualization methods in relation to their sensitivity/efficiency according to our insights. *Table 5.3* also provides an example of how to deal with primary antibody dilution in relation to the chromogen used. See also the primary antibody dilutions in the protocols of *Plate 8 G/H* and *I/J* in *Appendix H.1.3*.

   It is obvious that whenever a highly sensitive/efficient detection system must be employed for a single-staining method, this should be maintained for double staining.

5.2.2 Avoid the use of streptavidin–biotin techniques when the tissue of interest contains endogenous biotin

Many mammalian tissues contain biotin as a natural component. Hence, the use of (strept)avidin–biotin techniques carries the risk of intrinsic

Table 5.2. Relative sensitivity/efficiency of immunohistochemical detection systems

Very strong	Strong	Moderate	Weak
3-step StrepABComplex	**3-step** Strep/Enzyme via anti-FITC via anti-DIG APAAP or PAP	**3-step** RAM/Enzyme + GAR/Enzyme	**2-step** GAM/Enzyme
	2-step EnVision^{TM}*		
	1-step EPOS^{TM}/HRP*		

*EnVision^{TM} and EPOS^{TM} are DAKO A/S products.

Table 5.3. Relative sensitivity/efficiency of some substrates/chromogen systems applicable for multiple staining. Below the arrow an example is given for adapting the primary antibody dilution in relation to the substrate/chromogen sensitivity/efficiency

IGS epi-polarization	IGS bright field				GAL (Bluo-GAL/ ironcyanide)	
HRP (TMB)	HRP (DAB)	HRP (AEC)	AP (New Fuchsin)	AP (Fast Red)	AP (Fast Blue)	GAL (X-GAL/ ironcyanide)
AP (NBT/BCIP)	AP (DAKO Fast Red)		AP (Vector Red) -fluorescence-		AP (Vector Red) -bright field-	
HIGH						LOW →

Primary antibody dilution

| 1:200 | 1:100 | | 1:50 | | 1:20 | |

(strept)avidin-binding activity. Indeed, this type of (unwanted) staining has been found to be abundant in acetone-fixed cryostat sections of liver, kidney and intestinal organs (Wood and Warnke, 1981). The best method for the suppression of intrinsic (strept)avidin-binding activity is the subsequent application of 0.1% (strept)avidin and 0.01% D-biotin for 20 min each, prior to the incubation with the primary antibody (Wood and Warnke, 1981). In tissues that have been formalin-fixed and paraffin-embedded intrinsic (strept)avidin-binding activity is no longer present (Hsu *et al.*, 1981).

Possible pitfalls with (strept)avidin–biotin detection systems:

- It is claimed that intrinsic (strept)avidin-binding reoccurs again in formalin-fixed and paraffin-embedded tissue sections after heat-induced antigen retrieval, especially when using Tris/EDTA buffer (Shi *et al.*, 1997; Vyberg and Nielsen, 1998).
- The binding of avidin–enzyme or avidin–fluorochrome conjugates to mast cell granules is reported to occur in formalin-fixed and paraffin-embedded tissues (Bussolati and Gugliotta, 1983; Tharp *et al.*, 1985).
- Van den Oord *et al.* (1989) described the observation of the binding of biotinylated secondary antibodies and biotinylated lectins to hepatitis B surface antigen, present in the membrane or cytoplasm of hepatitis B virus-infected hepatocytes, in formalin- or Bouin-fixed liver tissue.
- (Strept)avidin–biotin interaction has been found to be negatively influenced by sugars, especially mannose, in enzyme-linked assays (Houen and Hansen, 1997). Whether this type of inhibitory effect also plays a role in immunohistochemistry is as yet unknown.

5.2.3 Apply a chromogen which fits to the antigen/cell type to be detected

Some chromogens provide a rather diffuse staining accuracy (*Table 5.4*). Therefore, these chromogens are less suitable for revealing tiny cellular

Table 5.4. Relative sharpness of localization of some substrate/chromogen systems

DAB	New Fuchsin	Fast Red TR	Fast Blue BB	X-GAL/ironcyanide
AEC	NBT/BCIP	Vector Red		
DAKO Fast Red	TMB	Bluo-GAL/ironcyanide		

CRISP   DIFFUSE

➤

structures. For example, the small dendrites from the antigen present-ing cells in skin epithelium (Langerhans cells) cannot always be detected properly using X-GAL as chromogen.

5.2.4 *Draw the planned double-staining procedure schematically*

After making a final selection of two individual detection systems, it is highly recommended that one prepares a schematic drawing and a step-by-step protocol of the planned double-staining procedure; check the drawing and protocol extensively for possible cross-reactions!

5.3 Selecting antibody tracers and colour combinations

With the development of immunohistochemical double-staining tech-niques many tracer enzyme and colour combinations have been exploited (*Table 5.5*). It appears from our experience that the selection of enzyme tracer(s) and colour combination(s) will depend largely on the aim of the study, the structure of tissues and cell characteristics, the double-staining method chosen and personal preferences (e.g. colour-blindness!).

In *Table 5.5a* a number of combinations are shown using one tracer enzyme, which is revealed histochemically in two different colours. Obviously, this approach can be applied only with *sequential* double-staining methods. The combination of two different enzymatic activities, as shown in *Table 5.5b*, allows *sequentially*, as well as *simultaneously*, performed double-staining methods.

Of the many double-staining colour combinations, those are selected which have been shown to be successful. The characteristics of these seven colour combinations are given below. Examples of six different colour combinations, including the visualization of a mixed colour at sites of co-localization are shown in *Plate 8A–K*.

- Brown–Red (Malik and Daymon, 1982) (*Plate 8 A/B*)
  HRP activity in brown → use DAB for visualization
  AP activity in red → use Fast Red or New Fuchsin for visualization
  (i) Two distinctly localized reaction products, showing moderate to poor contrast.

Table 5.5. Immunoenzyme double-staining colour combinations

(a) One enzyme, two different substrates/chromogens (for sequential techniques only)

HRP	AP	Basic colours		Reference
DAB	4-CN	brown	blue-grey	Nakane, 1968
AEC	4-CN	red	blue/grey	Richer and Mir, 1984
DAB/Ni	DAB	blue/black	brown	Hsu and Soban, 1982
DAB/Fe	AEC	green	red	Nemes, 1987
DAB	TMB/stabilized	brown	black	Rye et al., 1984
DAB	naphthol/pyronin	brown	red/purple	Sofroniew and Schrell, 1982
AEC	DAB/Ni	red	black	Poletti et al., 1987
VIP	DAB/Ni	red/purple	black	Lanciego et al., 1997
Fast Blue	New Fuchsin	blue	red	Feller et al., 1983

(b) Two enzymes, two different substrates/chromogens (for sequential, simultaneous or multistep techniques)

HRP	AP	GOX	GAL	Basic colours		Reference
DAB	Fast Blue	—	—	brown	blue	Mason and Sammons, 1978
DAB	New Fuchsin or Fast Red	—	—	brown	red	Malik and Daymon, 1982
DAB	—	glucose/NBT	—	brown	purple/blue	Campbell and Bhatnagar, 1976
DAB	BCIP/NBT	—	—	brown	purple/blue	Wolber and Lloyd, 1988
AEC	BCIP/NBT	—	—	red	purple/blue	Wolber and Lloyd, 1988
DAB	—	—	X-GAL	brown	turquoise	Sakanaka et al., 1988
AEC	Fast Blue	—	—	red	blue	Boorsma, 1984
TMB	Fast Red	—	—	green	red	Van der Loos et al., 1988
—	New Fuchsin or Fast Red	—	X-GAL	red	turquoise	Van der Loos et al., 1987
—	New Fuchsin	—	Bluo-GAL	red	blue	Van der Loos, this handbook
DAB	magenta phosphate	—	—	brown	purple	Avivi et al., 1994

   (ii) Not suitable for observing a mixed colour at sites of co-localization, because of lack of a well-defined mixed colour (Boorsma, 1984).
  (iii) A nuclear counterstain in blue with haematoxylin is optional.
  (iv) The Fast Red reaction product is soluble in alcohol and xylene. Mount aqueously.
   (v) The New Fuchsin reaction product is slightly soluble in alcohol and xylene. With high staining intensities organic mounting is no problem, but with low staining intensities aqueous mounting is recommended.

- Brown–Purple/Blue (Wolber and Lloyd, 1988) (*Plate 8 C/D*)
  HRP activity in brown → use DAB for visualization
  AP activity in purple/blue → use NBT/BCIP for visualization
   (i) DAB and NBT/BCIP both provides a sharply localized reaction product.
  (ii) Good colour contrast; better than with brown–red.
  (iii) Because of the dark nature of the NBT/BCIP reaction product, this colour combination is sometimes less suitable for observing a mixed colour at sites of co-localization. For that purpose, the NBT/BCIP reaction should be monitored microscopically; the appearance of a DAB–NBT/BCIP mixed colour will be greyish.
  (iv) Maintaining the original purple/blue colour of the NBT/BCIP reaction product, mount aqueously because it is partly soluble in alcohol and xylene. After mounting with an organic solvent the original purple/blue colour is less intense and shifted towards blue, close to the colour of Fast Blue reaction product. This option is especially suitable for a permanent nuclear counterstain with Methylgreen (*Appendix B.1.8*) (*Plate 6*).

- Brown–Blue (Mason and Sammons, 1978) (*Plate 8 E/F*)
  HRP activity in brown → use DAB for visualization
  AP activity in blue → use Fast Blue BB for visualization
   (i) Good colour contrast; better than with brown–red.
  (ii) DAB provides a sharply localized reaction product.
  (iii) In comparison with the other chromogens mentioned here, Fast Blue BB provides a rather diffusely localized reaction product (*Table 5.4*).
  (iv) The reaction sensitivity/efficiency of Fast Blue BB is certainly less than with NBT/BCIP (*Table 5.3*).
   (v) With restrictions, suitable for observing a mixed colour at sites of co-localization (greyish mixed colour) (Boorsma, 1984).
  (vi) A nuclear counterstain with Methylgreen (*Appendix B.1.8*) is optional but partly fades away after aqueous mounting.
  (vii) The Fast Blue BB reaction product is soluble in alcohol and xylene. Mount aqueously.

- Red–Blue (Boorma, 1984) *(Plate 8 G/H)*
  HRP activity in red → use AEC for visualization
  AP activity in blue → use Fast Blue BB for visualization
    (i) Very good colour contrast; better than brown–red and brown–blue.
   (ii) AEC provides a sharply localized reaction product, Fast Blue BB somewhat less *(Table 5.4)*.
  (iii) Be aware of the rather low staining sensitivity/efficiency of Fast Blue BB *(Table 5.3)*.
   (iv) Suitable for observing a mixed colour at sites of co-localization (purple mixed colour) (Boorsma, 1984).
    (v) A nuclear counterstain with Methylgreen *(Appendix B.1.8)* is optional but partly fades away after aqueous mounting.
   (vi) AP and HRP reaction products are soluble in alcohol and xylene. Mount aqueously. Store sections shielded from daylight.

- Red–Turquoise (Van der Loos *et al.*, 1987) *(Plate 8 I/J)*
  AP activity in red → use Fast Red or New Fuchsin for visualization
  GAL activity in turquoise → use X-GAL/ironcyanide for visualization
    (i) Superb colour contrast; even better than the red–blue combination.
   (ii) New Fuchsin/Fast Red both provide a sharply localized reaction product, X-GAL/ironcyanide is localized rather diffusely *(Table 5.4)*.
  (iii) Suitable for observing a mixed colour at sites of co-localization (purple/blue mixed colour).
   (iv) A nuclear counterstain with haematoxylin is optional, when two distinct cell populations are double-stained without a mixed colour at sites of co-localization.
    (v) AP reaction product with Fast Red is soluble in alcohol and xylene. Mount aqueously.
   (vi) The AP reaction product with New Fuchsin is slightly soluble in alcohol. Mount either organically (with some loss of AP reaction product) or aqueously.
  (vii) X-GAL may be replaced by Bluo-GAL (Aguzzi and Theuring, 1994), rendering a bluish reaction product, which is slightly more sensitive/efficient and more precisely localized *(Tables 5.3 and 5.4)*. Bluo-GAL combined with a red alkaline phosphatase reaction product may serve as an alternative red–blue colour combination; for example in cases of persistent endogenous peroxidase activity.

- Red–Green (Van der Loos *et al.*, 1988) (not shown)
  AP activity in red → use New Fuchsin for visualization
  HRP activity in green → use TMB for visualization
    (i) Suberb colour contrast. It serves as an alternative for the red–turquoise combination with AP and GAL, especially in those cases where GAL provides too little staining intensity *(Table 5.3)* or too diffuse staining accuracy *(Table 5.4)*.

(ii) New Fuchsin and tetramethylbenzidine (TMB) provide sharply localized reaction products.

(iii) Because of overstaining effects, TMB substrate is delicate concerning the dilution of the primary antibody; it needs at least a 10–50 × higher dilution than usual.

(iv) Suitable for observing a mixed colour at sites of co-localization (purple/blue mixed colour).

(v) A nuclear counterstain with haematoxylin is optional, when two distinct cell populations are double-stained without a mixed colour at sites of co-localization.

(vi) TMB reaction product is water soluble. Mount organically. Store sections shielded from daylight.

(vii) The AP reaction product with New Fuchsin is slightly soluble in alcohol. Mount organically (with some loss of AP reaction product).

- Red–Blue (double darkfield)
  (Van der Loos and Becker, 1994) (*Plate 8 K–M*)
  AP activity in red (epi-fluorescence) → use Vector Red for
                                          visualization
  IGS in blue (epi-polarization) → use Ultra Small Gold conjugates and
                                    silver enhancement

  (i) 0.8–1.0 nm ultra-small gold particles as antibody tracer will be enlarged by silver enhancement up to 50–80 nm. Applying a fluorescence microscope equipped with an epi-polarization filter block (or immunogold/silver block) the silver precipitate will be observed as blue/green spots. Fading of signal upon illumination as with 'classical' immunofluorescence is unknown with this observation method.

  (ii) The AP reaction product from Vector Red (Vector Labs) or CAS Red (Becton Dickinson) will show an intense red fluorescent signal when epi-illuminated with green light (rhodamine filter block). Fading of the signal upon illumination is minimal. It is estimated that fading is ~1000 × less compared with fading of FITC-fluorescence (Speel *et al.*, 1992).

  (iii) With epi-polarization microscopy the AP reaction product is invisible. With epi-fluorescence microscopy the silver precipitate is invisible. Both reaction products can thus be observed and recorded individually.

  (iv) When applying photographic double exposure a yellowish/white mixed colour at sites of co-localization can be recognized (*Plates 8K* and *10*).

  (v) The silver precipitate provides a granular sharply localized reaction product, the localization of the fluorescent AP reaction product is moderately accurate (*Table 5.4*). In contrast to Speel *et al.*, (1995) the use of Poly Vinyl Alcohol mixed with the AP chromogens was not found to improve the staining accuracy (Van der Loos, unpublished observation).

(vi) A nuclear counterstain with haematoxylin does not interfere with epi-polarization or epi-fluorescence.

(vii) Preparations can be stored at room temperature without any loss of epi-polarization or epi-fluorescence signal for at least 2–4 years.

(viii) Mount aqueously, preferably using glycergel (DAKO).

5.4 Double staining: possibilities for computerized imaging of two individual colours?

Of the colour combinations described above the turquoise–red shows the most superior contrast. We realized that this good contrast was because this colour combination is made up of almost opposing colours in the subtractive colour circle. For this reason we have chosen this colour combination to attempt computerized imaging. Two procedures have been tested:

1. Application of two narrow bandpass filters close to the optimum absorbance wavelengths of X-GAL and New Fuchsin reaction products (Lehr *et al.*, 1999). The red filter is 'dissolving' the red-stained macrophages almost completely (*Plate 7C*). However, the blue filter does not completely 'dissolve' the turquoise-stained smooth muscle cells (*Plate 7B*). Setting of a threshold grey value for those turquoise cells may result in the loss of specifically, but weakly, stained red cells. Tested blue filters with other nearby wavelengths did not appear to be effective either.

    The use of HRP, DAB (brown/yellow)/AP, Fast Blue (blue) as another optional colour combination for this purpose did not work either because of a wide absorbance spectrum for the DAB reaction product, which almost overlaps with the spectrum for the Fast Blue chromogen (Van der Loos, unpublished observation).

2. Lehr *et al.* (1997) recently described the use of relatively simple commercially available software (Photoshop version 3.0 or 4.0, Adobe Systems, Mountain View, CA, USA) for cellular imaging. With the aid of this software, specific colours can be easily selected, and mean density and pixel number can be read. We have found that this new imaging method works well for double-stained specimens with the turquoise/red combination applied (Lehr *et al.*, 1999). The pixel number of the turquoise and red areas, respectively, divided by the pixel number of the total area, is a direct measure for the surface areas taken by the two different colours (*Plates 7E* and *F*). As with the previous method, the disadvantage of losing weakly stained cells no longer exists. Also other colour combinations (brown–blue, red–blue) can be successful with this procedure (Lehr and Van der Loos, unpublished observations). *Plate 7A–F* shows an overview of both imaging methods applied for a low power micrograph.

6 Tips and trouble shooting

6.1 Which antigen, which colour?

From our practical experience we have found that the answer to this difficult question is highly dependent on tissue type, cellular composition, antigen/epitope distribution, personal preferences and, last but not least, the scientific aim. Nevertheless, some recommendations can be made:

- To detect mixed colours at sites of co-localization start testing the red–blue colour-combination.
- Two different cell populations can be demonstrated successfully by either applying the brown–red combination allowing a blue nuclear counterstain, or applying the turquoise–red combination, allowing computerized imaging (*Section 5.4*). The brown–blue, brown–purple/blue combinations may serve as alternatives.
- Regarding the antigen/epitope distribution in the tissue, the recommendations given in *Table 6.1* may serve as an initial guide. The idea behind this table is based on the natural colour densities of the different reaction products.

This initial guide (*Table 6.1*) is not given for the red–brown combination, since that combination consists of two almost equally densely coloured reaction products. The two reversed colour combinations should both be tested for each individual aim.

Table 6.1. Recommendations for colour combinations in relation with the antigen distribution in the tissue

	Red–turquoise	Brown–purple/blue	Brown–blue	Red–blue
Antigen present in the majority of cells, or antigen/epitope is strongly expressing → **stain in weakest colour:**	turquoise	brown*	brown*	blue
Antigen present in the minority of cells, or antigen/epitope is weakly expressing → **stain in darkest colour:**	red	purple/blue	blue	red

*Provided that the visualization with DAB is restricted to a moderate staining intensity, yielding a yellow/brown reaction product.

When a first double staining gives an unsatisfactory result, the reverse colour combination should always be tested. Spectacular differences between two reversed colour combinations are illustrated in *Plate 9A* and *B*!

6.2 Be aware of

6.2.1 Endogenous AP activity

Occurs in acetone-fixed cryostat sections, for example in vessels. AP enzyme tracer is generally derived from calf intestinal tissue, which contains a different AP isoenzyme than, for example, AP in the vessel wall. Add levamizole to inhibit all AP isoenzyme activities except for the intestinal AP activity (Borgers, 1973). It is reported that too high concentrations of levamisole may inhibit the activity of the AP label as well (Thisted, 1995). Because endogenous AP activity does not survive the fixation/embedding procedure, levamisole can be omitted for formalin-fixed paraffin sections.

6.2.2 Endogenous peroxidase activity

Sometimes regular blocking methods on cryostat sections may fail when applying 0.1% sodium azide + 0.3% peroxide in washing buffer (Li *et al.*, 1987). In such cases one should also test the more efficient blocking method with methanol + 0.3% peroxide, but one needs to be aware of its deleterious effects on many antigens and epitopes.

6.2.3 Endogenous biotin

Occurs in acetone-fixed cryostat sections, particularly of liver and kidney, and many other epithelia and cell types. It is reported to reoccur in paraffin sections after antigen retrieval (Shi *et al.*, 1997). An endogenous biotin blocking kit should be applied or detection methods used without biotin–streptavidin interaction, for example APAAP, DAKO-EPOS™, DAKO-EnVision™.

6.2.4 Brown tissue pigments

Melanin, iron pigment after haemorrhages, ceroid pigment (oxidized lipoproteins) in macrophages, lipofuscin in cardiac myocytes. In such cases the use of DAB is to be avoided (Van der Loos, 1998).

6.3 When the staining intensity in double staining is too low

Be aware of the fact that a once-established primary antibody dilution for single staining is usually based on the performance with, for example, a three-step streptavidin–biotin complex technique with DAB as peroxidase chromogen. This primary antibody dilution needs to be adapted when performing double staining with other detection systems and/or chromogens. In particular GAL activity revealed with X-GAL and AP activity revealed with Fast Blue BB, give a rather low sensitivity/ efficiency compared with NBT/BCIP, AEC or DAB substrates (*Table 5.3*) (De Jong *et al.*, 1985; Scopsi and Larsson, 1986; Trojanowski *et al.*, 1983). Therefore, this lower sensitivity/efficiency should be compensated for by using either a 2–4 × higher concentration of the primary antibody, a more efficient/sensitive detection method (*Table 5.2*), using overnight incubation at 4°C instead of 60 min at room temperature, or combinations of these actions. An example of adapting the primary antibody concentration with respect to the chromogens used, is illustrated in the detailed protocols for *Plate 8* in *Appendix H.1.3*.

6.4 Order of enzymatic development

For the *sequential* double-staining technique, it is strictly recommended that the HRP activity using DAB as chromogen is developed first. Because this chromogen has the potential for sheltering the first set of (immuno)-reagents, DAB reaction product contributes to the prevention of cross-reactions between the first and second staining sequence (see *Section 3.1*). Unlike the DAB reaction product, AEC and AP or GAL reaction products do not have this sheltering characteristic (Van den Brink *et al.*, 1990).

For simultaneous or multi-step double-staining techniques applying HRP and AP as enzyme labels, AP activity is generally developed first, and, after a brief wash, HRP activity as second. It is thought that the peroxide in the HRP visualization medium may damage the AP activity (Malik and Daymon, 1982). To maintain 'good laboratory practice' we usually keep to this order of development. However, we have seen that in case of low HRP activity and strong AP activity a reversed development procedure may improve the staining quality without any problems concerning the AP staining intensity (Van der Loos, unpublished observation).

For the GAL/AP combination there seems to be no preference regarding which of the two substrates is used first (Van der Loos, unpublished observation). Again, for reasons of 'good laboratory practice' we have chosen to develop GAL activity first, and AP second.

- *Tip 1*. Initially, it is important not to handle too many slides in one experiment, particularly when applying more than one antibody combination, or more than one colour combination.
- *Tip 2*. The red–blue combination, as obtained from HRP/AEC and AP/ Fast Blue; is a good colour combination to begin.
- *Tip 3*. Try to imagine how your tissue specimen will look after double staining. This may help in determining the appropriate colour combination.

6.5 Common mistakes with double staining

The most common mistakes in double-staining protocols are best illustrated by trouble shooting protocols 1–6, which have one or more mistakes (washing steps have been omitted from the protocols). Please try to locate those mistakes, and make suggestions to correct the protocols. The answers can be found in *Appendix H.1.2.*

TROUBLE SHOOTING No. 1

Tissue: human
antigens 1 and 2 are present in different cell types

Normal goat serum		10 min
{ Mouse antibody 1 { Rabbit antibody 2	(cocktail)	60 min
{ Rabbit anti-mouse/HRP { Swine anti-rabbit/AP	(cocktail)	30 min

AP detection in blue
HRP detection in red

Results:
antigen 1:   specific purple staining (red was expected)
antigen 2:   specific purple staining (blue was expected)

TROUBLE SHOOTING No. 2

Tissue: rat
antigens 1 and 2 are present in different cell types

Antibodies: GAM/HRP and GAR/AP are secondary reagents, designed for use on human tissues.

Normal goat serum		10 min
{ Mouse antibody 1 { Rabbit antibody 2	(cocktail)	60 min
{ Goat anti-mouse/HRP { Goat anti-rabbit/AP	(cocktail)	30 min

AP detection in blue
HRP detection in red

Results:
antigen 1:   specific red staining, with strong non-specific
background staining
antigen 2:   specific blue staining, with moderate non-specific
background staining

TROUBLE SHOOTING No. 3

| **Tissue:** | human |
| | antigens 1 and 2 are present in different cell types |

Normal goat serum	10 min
Mouse antibody 1	60 min
Rabbit anti-mouse/biotin	30 min
{ Streptavidin/AP { Normal mouse serum (cocktail)	30 min
Mouse antibody 2/FITC	60 min
Rabbit anti-FITC	15 min
Goat anti-rabbit/HRP	30 min

AP detection in blue
HRP detection in red

Results:

antigen 1:	specific purple staining (blue was expected)
antigen 2:	specific red staining

TROUBLE SHOOTING No. 4

| **Tissue:** | human |
| | antigens 1 and 2 are present in different cell types |

Normal goat serum	10 min
{ Mouse antibody 1/Phycoerythrin { Mouse antibody 2/FITC (cocktail)	60 min
{ Rabbit anti-Phycoerythrin { Goat anti-FITC (cocktail)	30 min
{ Swine anti-rabbit/AP { Swine anti-goat/HRP (cocktail)	30 min

AP detection in blue
HRP detection in red

Results:

antigen 1:	specific blue staining
antigen 2:	specific red staining, with heavy non-specific background staining

TROUBLE SHOOTING No. 5

Tissue:	human
	antigens 1 and 2 are present in different cell types

Normal goat serum	10 min
Mouse antibody 1	60 min
Goat anti-mouse/HRP	30 min
Swine anti-goat/HRP (for enhancement of GAM/HRP)	30 min
Normal mouse serum	30 min
Mouse antibody 2/FITC	60 min
Rabbit anti-FITC	15 min
Goat anti-rabbit/AP	30 min

AP detection in blue
HRP detection in red

Results:

antigen 1:	specific purple staining with heavy non-specific background staining (red was expected)
antigen 2:	specific blue staining

TROUBLE SHOOTING No. 6

Tissue:	human, no endogenous biotin
	antigens 1 and 2 are present in different cell types

Normal goat serum		10 min
Mouse antibody 1		60 min
Goat anti-mouse/biotin		30 min
{ Streptavidin-biotin complex/HRP Normal mouse serum	(cocktail)	30 min
d-biotin	0.01%	15 min
Mouse antibody 2/biotin		60 min
Streptavidin/AP		30 min

AP detection in blue
HRP detection in red

Results:

antigen 1:	specific purple staining (red was expected)
antigen 2:	specific blue staining

6.6 Restrictions for the visualization of a mixed colour at sites of co-localization using two enzymatic activities

There are some restrictions for a feasible observation of a mixed colour at sites of co-localization using enzyme tracers:

- When the two antigens under study are present in highly varying amounts, a subtle mixed colour will be easily missed. In such situations 'classical' double immunofluorescence supported by confocal laser scanning microscopy, is recommended (Brandtzaeg *et al.*, 1997). Alternatively, the combination of an IGS technique and a (non-fading) fluorescent AP staining method can also be tested (Van der Loos and Becker, 1994). This latter technique is illustrated in *Plate 10*, showing the co-localization of a low-expressing cytokine and abundantly present cell-structural marker. Moreover, this IGS/AP combination can also be observed using confocal laser scanning microscopy (Van der Loos, unpublished observation).

- It is generally believed that even the best enzymatic reaction products are less accurate than fluorochromes. Particularly in densely clustered cellular areas, where a poor staining accuracy of one of the two reaction products will lead to false colour mixing. As shown in the six colour combination characteristics with two enzymatic activities in *Section 5.3*, most have one rather diffusely localized chromogen involved. With one combination both reaction products are highly accurate, but there is a poor colour contrast (red–brown). If none of these six colour combinations is suitable, double-immunofluorescence techniques or the combination of an IGS technique with a fluorescent AP staining method should again be chosen. The accuracy of an IGS technique is thought to be close to that of immunofluorescence.

- In a case of two low abundance antigens/epitopes, which can be barely detected using highly efficient/sensitive single-immunohistochemical techniques, double staining might be almost impossible. Since the sensitivity/efficiency of immunofluorescence and immunohistochemical techniques are almost equal, double staining with two fluorochromes might also be a problem. In such instances, advanced microscopic systems with the possibility of electronic enhancing the fluorescence images can be of great help.

7 Double-staining speciality: combination of immunohistochemistry with TUNEL

The terminal transferase dUTP nick end labelling (TUNEL) technique has become the standard for identifying DNA fragmentation, a characteristic phenomenon for apoptotic nuclei during their final 'execution' phase of programmed cell death. Because the TUNEL technique is known to inherit a number of pitfalls it should be emphasized that beside TUNEL positivity, other criteria should be available for the definitive recognition of an apoptotic nucleus (Kockx, 1998).

Frequently, investigators are interested in which type of cell the apoptotic cell death occurs. It appears in the literature that there is not very much consensus on a generally applicable double-staining procedure. Most papers deal with a sequential enzymatic procedure, starting with a complete TUNEL staining, followed by immunohistochemistry. Others followed a reversed sequential approach beginning with complete immunohistochemistry, followed by TUNEL. In my opinion, all enzymatic procedures to date have yielded poor results with unconvincing pictures. Only a few applications using a double fluorescence technique imaged with a confocal laser scanning microscope system have given satisfactory results (e.g. Ikeda *et al.*, 1997).

7.1 Sequence of application

The combination of immunohistochemistry with TUNEL in one tissue specimen may be considered as a special double-staining technique, following the *sequential* staining concept. Surprisingly, none of the applications to date have paid any attention to shielding problems by enzymatic reaction products as described for the DAB reaction product with the *sequential* staining technique of two antigens (*Section 3.1*).

59

Moreover, no publication has referred to shielding problems as encountered with the combination of *in situ* hybridization and immunohistochemistry (Robben *et al.*, 1994; Speel *et al.*, 1995 in review; Van den Brink *et al.*, 1990; Van der Loos *et al.*, 1989b), a technique which is after all highly similar to TUNEL/immunohistochemistry double staining. Furthermore, none of the applications mentioned the use of enzymatic or heat-induced antigen retrieval pretreatment for immunohistochemistry, or the testing of an optimal colour contrast. Therefore, we have initiated a series of experiments with the objective of elucidating a generally applicable protocol for the TUNEL/immunohistochemistry double-staining specialty, answering such questions as: which is the best order of application of both techniques, and which is the best chromogen/colour combination?

Based on the TUNEL protocol according to Gavrieli *et al.* (1992) including a protease K tissue pretreatment preceding TUNEL, three different approaches for the combination of TUNEL and immunohistochemistry were tested on tonsils that had been formalin fixed immediately after surgery. The following advantages/disadvantages were recorded.

7.1.1 *First immunohistochemistry, including enzymatic visualization, followed by TUNEL, including enzymatic visualization*

Shielding effects by the reaction products obtained from immunohistochemistry resulted in a negative or weak TUNEL signal. This shielding effect was found to be minimal with the chromogens AEC and X-GAL for HRP and GAL activity, respectively. Although heat-induced antigen retrieval with citrate pH 6.0, in combination with protease K digestion, is suggested as a tissue pretreatment for TUNEL (Sträter *et al.*, 1995), we obtained poor results with double-staining procedures starting with a primary antibody requiring heat-induced antigen retrieval as a pretreatment. For this sequential approach starting with complete immunohistochemistry, we obtained good double-staining results only with a primary antibody which either does not require a tissue pretreatment or requires only proteolytic pretreatment. The enzymatic digestion serves both as a tissue pretreatment for TUNEL and immunohistochemistry. However, some antigens/epitopes have proven vulnerable to the enzymatic digestion procedure with protease K prior to the TUNEL. For example we have found a weak/negative result with anti-smooth muscle α-actin, clone 1A4 after enzymatic digestion with pepsin.

7.1.2 *First TUNEL, including enzymatic visualization, followed by immunohistochemistry, including enzymatic visualization*

In general, this order of performance gave better results than the one in *Section 7.1.1*. The main problem is sometimes a weak immunohistochemical signal due to antigen/epitope shielding effects by the TUNEL reaction products. Although the DAB reaction product was only restricted to apoptotic nuclei, the following immunohistochemical staining intensity was overall moderately to severely decreased. This problem was also reported for the combination of immunohistochemistry and *in situ* hybridization (Van den Brink *et al.*, 1990). When using chromogens other than DAB, we encountered the problem that some reaction products, for example NBT/BCIP, were not able to resist the heat-induced antigen retrieval step.

7.1.3 First TUNEL reaction, without any further detection, followed by immunohistochemistry simultaneously with appropriate detection for TUNEL. Ending up with two subsequent enzymatic visualizations

Because of the shielding problems mentioned earlier, we tested the so far unexploited option of performing TUNEL first, then immunohistochemistry simultaneously performed with the detection of the digoxigenin label and finally visualization of the two enzymatic activities. For the order of enzymatic development one should stick to suggestions made in *Section 6.4*. This TUNEL/immunohistochemistry double-staining procedure resulted in superior results compared with the two approaches above. Heat-induced antigen retrieval, needed for the detection of a particular antigen/epitope, could be performed after TUNEL. It was observed that the inbuilt dUTP-DIG was kept exactly at its original location, although, the original TUNEL staining intensity was slightly decreased. This approach is technically specified in a 'blueprint' protocol in *Appendix D.1.7*. Typical examples of this procedure without and with heat-induced antigen retrieval are shown in *Plates 11* and *12*, respectively.

7.2 Colour combinations

As a consequence of the above, a number of colour combinations successful in immunohistochemical double staining are ruled out for combinations with TUNEL. For example, the use of DAB as chromogen for HRP activity should be avoided for the first staining sequence in the approaches given in *Sections 7.1.1* and *7.1.2*. Because double staining with TUNEL and immunohistochemistry for cytoplasmic and membrane markers and TUNEL does not lead to co-localization including colour mixing, alternative colour combinations can be tested. For example we have successfully combined AEC in red for immunohistochemistry and BCIP/NBT in blue/purple for TUNEL when performing the approach given in *Section 7.1.1*.

Most successful was the approach given in *Section 7.1.3*, staining TUNEL in purple/blue with NBT/BCIP and immunohistochemistry with

DAB in brown. This combination leaves the possibility of a nuclear counterstain with Methylgreen (*Plate 11*). However, this purple/blue–brown colour combination gave poor results when the TUNEL-signal was located inside the starry sky macrophages which have scavenged the apoptotic B cells. For that purpose, the brown–red colour combination resulted unexpectedly in a better contrast (*Plate 12*).

8 Immunoenzyme triple staining

For immunoenzyme triple staining three enzymatic activities of HRP, AP and GAL need to be combined in one experiment. As stressed previously for double staining, the prevention of cross-reaction in both *simultaneous* and *sequential* detection procedures, is of major importance for a successful multiple-staining procedure. This is certainly the case with triple staining, and the problems of overcoming it are even more prominent. In general, triple staining consists of a combination of two double-staining concepts (dependent on the nature of the primary antibodies), or a combination of a double-staining and an aberrant single-staining procedure.

The following combinations we have found to work well:

- *indirect/indirect/indirect* with three monoclonal antibodies of different Ig isotypes or IgG subclasses (Marijianowski *et al.*, 1995);
- *indirect/indirect/indirect* with two monoclonal antibodies of different Ig isotypes or IgG subclasses and one rabbit polyclonal antibody (*Plate 13*);
- *indirect/indirect/indirect* with three antibodies of different animal species;
- multi-step *indirect/direct* combined with a third *indirect* procedure (*Plate 14*);
- multi-step *indirect/direct* combined with a third *direct* procedure (Van der Loos *et al.*, 1987, 1989a, 1993);
- combination of any simultaneous double-staining concept with *sequential* double staining, starting with HRP/DAB, followed by microwave treatment to remove the first set of reagents (Van den Brink *et al.*, 1999) (*Plate 15A/B*);
- combination of any double staining procedure with a lectin staining.

8.1 Triple staining without co-localization(s)

The most important application of immunoenzyme triple staining is to demonstrate three different cell types and/or cell constituents in a single

63

tissue specimen. It may provide a superb overview of the relationship between the cells and/or cell constituents, which might be easily missed in three serial sections. Most authors using triple staining in this regard had developed their own system strictly tailored to their own primary antibodies and application. The examples shown in *Plates 13–15* provide triple staining marking three different cell constituents, using commercially available reagents only. Detailed protocols as applied for *Plates 13–15* can be found in *Appendix H.1.3*.

8.2 Triple staining with co-localization(s)

As mentioned in *Section 4.1*, triple staining might be used to verify what is already evaluated from double-staining results. It is important, however, to realize that under optimal conditions co-localization of two antigens can be visualized with immunoenzyme double staining, whereas it will be much more difficult, or even impossible, to discriminate between double- and triple-stained cell structures obtained from triple-staining experiments. For example, a mixed colour from GAL/X-GAL (turquoise) and AP/Fast Blue (blue) would be very obscure (Van der Loos *et al.*, 1989a). For a more definitive elucidation of double- or triple-stained cells, it might be helpful to record photographically the same tissue spot, after each subsequent enzyme activity visualization (Van der Loos *et al.*, 1993). It is obvious that this laborious procedure can only be done with one or two tissue sections. For larger series of tissue specimens, fluorescence techniques will be superior to the enzymatic methods (Brandtzaeg *et al.*, 1997).

8.3 Triple-staining colour combinations and the order of visualization

Immunoenzyme triple-staining procedures have been rarely applied and published. Van Noorden *et al.* (1986) and Claasen *et al.* (1986) described a triple-staining procedure for marking three different cellular constituents using only two enzymatic activities. These procedures were based on the occurrence of a mixed reaction product applied as the third colour.

Nakane (1968), Kiss *et al.* (1988) and Lanciego *et al.* (1997) described triple-staining procedures of a sequential concept based on three different HRP chromogens.

Hopman *et al.* (1997) used multi-target DNA *in situ* hybridization to visualize three different chromosomes stained with HRP/DAB in brown, AP/New Fuchsin in red and HRP/TMB in green.

Recently, a series of 72 patient biopsies were subjected to a 3 day immunoenzyme triple staining as performed by the Department of Gastroenterology, Academical Medical Centre, Amsterdam, The Netherlands (Van den Brink *et al.*, unpublished data). *Plate 15A/B* shows an example of this procedure, and the protocol can be found in *Appendix H.1.3*. To my knowledge this is the largest immunoenzyme triple staining series ever.

One should realize that the streptavidin–biotin interaction can be applied for only one antigen/epitope. Because of the poor availability of GAL-conjugated second step antibodies and the poor staining sensitivity/efficiency of GAL as a tracer enzyme in two-step detection procedures, a three-step detection via streptavidin–biotin is preferably applied to this enzyme. In case of strongly expressing antigens/epitopes a two-step procedure with GAL-conjugated anti-rabbit and anti-mouse reagents may work (Southern Biotechnology Associates, bulk order on special request!). The production of GAL-conjugated antibodies is outlined in *Appendix G.1.3*.

The most successful colour combination we have applied to immunoenzyme triple staining is GAL/X-GAL in turquoise, AP/Fast Blue in blue and HRP/AEC in red, developed in that order (Marijianowski *et al.*, 1995; Van der Loos *et al.*, 1987, 1989a, 1993). Following the recommendations in *Table 6.1* for double staining, the most abundantly/strongly present antigen should be stained in the weakest colour, turquoise, the moderately abundant antigen in blue and the antigen which is present least in red. This approach will give good results in many cases (*Plates 13* and *14*).

For triple staining involving X-GAL and AEC chromogens (Van der Loos *et al.*, 1993), we have seen a slight inhibitory effect of the X-GAL substrate solution on the HRP activity. The longer the incubation time for GAL activity visualization, the larger the decrease in HRP activity. Alternatively, starting with HRP/AEC showed a slight inhibitory effect on the GAL/X-GAL activity. The alternative order of visualization should be considered whenever the staining intensity with HRP is too weak in the normal staining order. Interference of enzymatic activities of this nature can sometimes be compensated for successfully by using a higher concentration of primary antibody, resulting in a shorter enzymatic activity development time.

Apart from the turquoise/blue/red colour combination, other combinations might be successful too. *Plate 15* shows a combination of HRP/DAB in brown/yellow, GAL/X-GAL in turquoise and AP/Fast Red in red (Van den Brink *et al.*, 1999). This colour combination even allowed a blue nuclear counterstain with haematoxylin, but is only suitable for detecting three different cell types without co-localization(s), because it does not allow a mixed colour between brown/yellow and turquoise.

Appendix A

References

Adams, J.C. (1992) Biotin amplification of biotin and horseradish peroxidase signal in histochemical stains. *J. Histochem. Cytochem.* **40:** 1457–1463.

Aguzzi, A. and Theuring, F. (1994) Improved in situ β-galactosidase staining for histological analysis of transgenic mice. *Histochemistry* **102:** 477–481.

Avivi, C., Rosen, O. and Goldstein, R.S. (1994) New chromogens for alkaline phosphatase histochemistry: salmon and magenta phosphate are useful for single- and double-label immunohistochemistry. *J. Histochem. Cytochem.* **42:** 551–554.

Avrameas, S. (1969a) Coupling of enzymes with glutaraldehyde. Use of conjugates for the detection of antigens and antibodies. *Immunochemistry* **6:** 43–52.

Avrameas, S. (1969b) Indirect immunoenzyme techniques for the intracellular detection of antigens. *Immunochemistry* **6:** 825–831.

Battifora, H. and Kopinski, M. (1986) The influence of protease digestion and duration of fixation on the immunostaining of keratins. A comparison of formalin and ethanol fixation. *J. Histochem. Cytochem.* **34:** 1095–1100.

Battifora, H. (1991) Assessment of antigen damage in immunohistochemistry. *Am. J. Clin. Pathol.* **96:** 669–671.

Behringer, D.M., Meyer, K-H. and Vey, R.W. (1991) Antibodies against neuroactive amino acids and neuropeptides II. Simultaneous immunoenzymatic double staining with labeled primary antibodies of the same species and a combination of the ABC method and the hapten-anti-hapten bridge (HAB) technique. *J. Histochem. Cytochem.* **39:** 761–770.

Bisgaard, K., Lihme, A., Rolsted, H. and Pluzek, K.-J. (1993) Polymeric conjugated for enhanced signal generation in enzyme immuno assays. Abstract. Scandinavian Society for Immunology XXIVth Annual Meeting, Aarhus, Denmark.

Bisgaard, K. and Pluzek, K.-J. (1996) Water soluble polymer conjugates for enzyme immuno assays. Short communication. 10th International Congress of Histochemistry and Cytochemistry, Kyoto, Japan.

Bobrow, M.N., Harris, T.D., Shaughnessy, K.J. and Litt, G.J. (1989) Catalyzed reporter deposition, a novel method of signal amplification: Application to immunoassays. *J. Immunol. Meth.* **125:** 279–285.

Bondi, A., Chieregatti, G., Eusebi, V., Fulcheri, E. and Bussolati, G. (1982) The use of β-galactosidase as a tracer in immunohistochemistry. *Histochemistry* **76:** 153–158.

Boorsma, D.M. (1983) Preparation of HRP-labelled antibodies, in *Immunohistochemistry IBRO Handbook* (A.C. Cuello ed.), pp. 87–100. John Wiley and Sons, Chichester, UK.

Boorsma, D.M. (1984) Direct immunoenzyme double staining applicable for monoclonal antibodies. *Histochemistry* **80:** 103–106.

Borgers, M. (1973) The cytochemical application of new potent inhibitors of alkaline phosphatase. *J. Histochem. Cytochem.* **21:** 812–824.

Bosman, F.T., Cramer-Knijnenburg, G. and Bergen Henegouw, J. (1983) Efficiency and sensitivity of indirect immunoperoxidase methods. *Histochemistry* **77**: 185–194.

Brandtzaeg, P. (1982) Tissue preparation methods for immunohistochemistry, in: *Techniques in immunohistochemistry*, Vol. 1 (GR Bullock, and P Petrusz, eds), pp. 1–75, Academic Press, London.

Brandtzaeg, P. (1998) The increasing power of immunohistochemistry and cytochemistry. Historical review with a personal touch. *J. Immunol. Meth.* **216**: 49–67.

Brandtzaeg, P., Halstensen, T.S., Huitfeldt, H.S. and Valnes, K.N. (1997) Immunofluorescence and immunoenzyme histochemistry, in *Immunochemistry 2-A Practical Approach* (A.P. Johnstone and M.W. Turner eds), pp. 71–130. Oxford University Press, Oxford, UK.

Brandtzaeg, P. and Rognum, T.O. (1983) Evaluation of tissue preparation methods and paired immunofluorescence staining for immunocytochemistry of lymphomas. *Histochem. J.* **15**: 655–689.

Breitschopf, J. and Suchanek, G. (1996) Detection of mRNA on paraffin embedded material of the central nervous system with DIG-labeled RNA probes, in: *Non-Radioactive in situ Hybridization*. Application Manual, 2nd edn, pp. 136–14. Boehringer, Mannheim, Germany.

Buckel, P. and Zehelein, E. (1981) Expression of *Pseudomonas fluorescens* D-galactose dehydrogenase in *E. coli. Gene* **16**: 149–159.

Burnstone, M.S. (1961) Histochemical demonstration of phosphatases in frozen sections with naphthol AS-phosphates. *J. Histochem. Cytochem.* **9**: 146–153.

Bussolati, G. and Gugliotta, P. (1983) Nonspecific staining of mast cells by avidin–biotin complexes (ABC). *J. Histochem. Cytochem.* **31**: 1419–1421.

Campbell G.T. and Bhatnagar A.S. (1976) Simultaneous visualization by light microscopy of two pituitary hormones in a single tissue section using a combination of indirect immunohistochemical methods. *J. Histochem. Cytochem.* **24**: 448–452.

Cattoretti, G., Pileri, S., Parravicini, C., Becker, M.H.G., Poggi, S., Bifulco, C., Key, G., D'Amato, L., Sabattini, E., Feudale, E., Reynolds, F., Gerdes, J. and Rilke, F. (1993) Antigen unmasking on formalin-fixed, paraffin-embedded tissue sections. *J. Pathol.* **171**: 83–98.

Coggi, G., Dell'Orth, P. and Viale, G. (1986) Avidin–biotin methods, in *Immunocytochemistry. Modern Methods and Applications* (J.M. Polak and S. Van Noorden eds), 2nd edn, pp. 54–70. Wright, Bristol, UK.

Coons, A.H., Creech, H.J. and Jones, R.N. (1941) Immunological properties of an antibody containing a fluorescent group. *Proc. Soc. Exp. Biol. Med.* **47**: 200–202.

Coons, A.H. and Kaplan, M.H. (1950) Localization of antigen in tissue cells. II. Improvements in a method for the detection of antigens by means of fluorescent antibody. *J. Exp. Med.* **91**: 1–13.

Cordell, J.L., Falini, B., Erber, W.N., Ghosh, A.K., Abdulaziz, Z., MacDonald, S., Pulford, K.A.F., Stein, H. and Mason, D.Y. (1984) Immunoenzymatic labeling of monoclonal antibodies using immune complexes of alkaline phosphatase and monoclonal anti-alkaline phosphatase (APAAP complexes). *J. Histochem. Cytochem.* **32**: 219–229.

Chaubert, P., Bertholet, M-M., Correvon, M., Laurini, S. and Bosman, F.T. (1997) Simultaneous double immunoenzymatic labeling: a new procedure for the histopathological routine. *Modern. Pathol.* **10**: 585–591.

Claasen, E., Boorsma, D.M., Kors, N. and Van Rooijen, N. (1986) Double-enzyme conjugates, producing an intermediate colour, for simultaneous and direct detection of three different intracellular immunoglobulin determinants with only two enzymes. *J. Histochem. Cytochem.* **34**: 423–428.

Curran, R.C. and Jones, E.L. (1977) Immunoglobulin-containing cells in human tonsils as demonstrated by immunohistochemistry. *Clin. Exp. Immunol.* **28**: 103–115.

DAKO Guide (1997) *A Guide to Demasking of Antigen on Formalin-Fixed, Paraffin-Embedded Tissue*. DAKO Handbook 00091. DAKO A/S, Glostrup, Denmark.

Danscher, G. (1981) Localization of gold in biological tissue. A photochemical method for light and electron microscopy. *Histochemistry* **71**: 81–88.

De Boer, O.J., Hirsch, F., Van der Wal, A.C., Van der Loos, C.M., Das, P.K. and Becker, A.E. (1997) Co-stimulatory molecules in human atherosclerotic plaques: an indication of antigen specific T-lymphocyte activation. *Atherosclerosis* **133:** 227–234.

De Bruijn, E.M.C.A. (1992) Formaldehyde fixation in immunohistochemistry (abstract). *Histochem. J.* **24:** 493.

De Jong, A.S.H., Van Kessel-Van Vark, M. and Raap, A.K. (1985) Sensitivity of various visualization methods for peroxidase and alkaline phosphatase activity in immunoenzyme histochemistry. *Histochem. J.* **17:** 1119–1130.

DeLellis, R.A. and Kwan, P. (1988) Technical considerations in the immunohistochemical demonstration of intermediate filaments. *Am. J. Surg. Pathol.* **12** (Suppl. 1): 17–23.

DeLellis, R.A., Sternberger, L.A., Mann, R.B., Banks, P.M. and Nakane, P.K. (1979) Immunoperoxidase techniques in diagnostic pathology. *Am. J. Pathol.* **71:** 483–488.

Denk, H., Radaszkiewicz, T. and Weirich, E. (1977) Pronase pretreatment of tissue sections enhances sensitivity of the unlabelled antibody-enzyme (PAP) technique. *J. Immunol. Meth.* **15:** 163–167.

De Waele, M., Renmans, W., Segers, E., Jochmans, K. and Van Kamp, B. (1988) Sensitive detection of immunogold-silver staining with darkfield and epi-polarization microscopy. *J. Histochem. Cytochem.* **36:** 679–683.

Evers, P. and Uylings, H.B. (1994) Microwave-stimulated antigen retrieval is pH and temperature dependent. *J. Histochem. Cytochem.* **42:** 1555–1563.

Falini, B., Abdulaziz, Z., Gerdes, J., Canino, S., Ciani, C., Cordell, J., Knight, P.M., Stein, H., Grignani, F., Martelli, M.F. and Mason, D.Y. (1986) Description of a sequential staining procedure for double immuno-enzymatic staining of pairs of antigens using monoclonal antibodies. *J. Immunol. Meth.* **93:** 265–273.

Falini, B., De Solas, I., Halverson, C., Parker, J.W. and Taylor, C.R. (1982) Double labeled-antigen method for demonstration of intracellular antigens in paraffin-embedded tissues. *J. Histochem. Cytochem.* **30:** 21–26.

Falini, B. and Taylor, C.R. (1983) New developments in immunoperoxidase techniques and their application. *Arch. Pathol. Lab. Med.* **107:** 105–117.

Farmilo, A.J and Stead, R.H. (1989) Fixation in immunocytochemistry. In: *Immunochemical staining methods.* DAKO Handbook (S.J. Naish ed.), pp. 24–29. DAKO Corporation, Carpinteria, CA.

Faulk, W. and Taylor, G. (1971) An immunogold colloid method for the electron microscope. *Immunochemistry* **8:** 1081–1083.

Feller, A.C., Parwaresch, M.R., Wacker, H.H., Radzun, H-J. and Lennert, K. (1983) Combined immunohistochemical staining for surface IgD and T-lymphocyte subsets with monoclonal antibodies in human tonsils. *Histochem. J.* **15:** 557–562.

Gavrieli, Y., Sherman, Y. and Ben-Sasson, S.A. (1992) Identification of programmed cell death *in situ* via specific labeling of nuclear DNA fragmentation. *J. Cell. Biol.* **119:** 493–501.

Geuze, H.J., Slot, J.W., Van der Ley, P., Schuer, R. and Griffith, J. (1981) Use of colloidal gold particles in double labeling immuno electron microscopy on ultrathin frozen sections. *J. Cell. Biol.* **89:** 653–665.

Ghandour, M.S., Langley, O.K., Vincendon, G. and Gombos, G. (1979) Double labeling immunohistochemical technique provides evidence of the specificity of glial cell markers. *J. Histochem. Cytochem.* **27:** 1634–1637.

Gillitzer, R., Berger, R. and Moll, H. (1990) A reliable method for simultaneous demonstration of two antigens using a novel combination of immunogold–silver staining and immunoenzymatic labeling. *J. Histochem. Cytochem.* **38:** 307–313.

Glezerov, V. (1986) Simultaneous detection of two lymphocyte surface antigens: combination of indirect and direct immunofluorescence methods with monoclonal antibodies. *J. Histotechnol.* **9:** 15–16.

Graham, R.C. and Karnovsky, M.J. (1966) The early stages of absorption of injected horseradish peroxidase in the proximal tubules of mouse kidney: ultra structural cytochemistry by a new technique. *J. Histochem. Cytochem.* **14:** 291–302.

Graham, R.C., Lundholm, U. and Karnovsky, M.J. (1965) Cytochemical demonstration of peroxidase activity with 3-amino-9-ethylcarbazole. *J. Histochem. Cytochem.* **13:** 150–152.

Guesdon, J-L., Ternnck, T. and Avrameas, S. (1979) The use of avidin–biotin interaction in immunoenzymatic techniques. *J. Histochem. Cytochem.* **27:** 1131–1139.

Hall, P.A., Stearn, P.M., Butler, M.G. and D'Ardenne, A.J. (1987) Acetone/periodate-lysine-paraformaldehyde (PLP) fixation and improved morphology of cryostat sections for immunohistochemistry. *Histopathology* **11:** 93–101.

Hancock, W.W., Becker, G.J. and Atkins, R.C. (1982) A comparison of fixatives and immunohistochemical techniques for use with monoclonal antibodies to cell surface antigens. *Am. J. Clin. Pathol.* **78:** 825–831.

Heras, A., Roach, C.M. and Key, M.E. (1995) Enhanced polymer detection system for immunohistochemistry (abstract). *Lab. Invest.* **72:** 165A.

Holgate, C.S., Jackson, P., Cowen, P.N. and Bird, C.C. (1983) Immunogold-silver staining: a new method of immunostaining with enhanced sensitivity. *J. Histochem. Cytochem.* **31**: 938–944.

Hopman, A.H.N., Claessen, S. and Speel, E.J.M. (1997) Multi-colour brightfield *in situ* hybridisation on tissue sections. *Histochem. Cell. Biol.* **108:** 291–298.

Houen, G. and Hansen, K. (1997) Interference of sugars with the binding of biotin to streptavidin and avidin. *J. Immunol. Meth.* **210:** 115–123.

Hsu, S-M., Raine, L. and Fanger, H. (1981) Use of avidin–biotin–peroxidase complex (ABC) in immunoperoxidase techniques: a comparison between ABC and unlabeled antibody (PAP) procedures. *J. Histochem. Cytochem.* **29:** 577–580.

Hsu, S-M. and Soban, E. (1982) Colour modification of diaminobenzidin (DAB) precipitation by metallic ions and its application for double immunohistochemistry. *J. Histochem. Cytochem.* **30:** 1079–1082.

Ikeda, H., Nakamura, Y., Hiwasa, T., Sakiyama, S., Kuida, K., Su, M.S. and Nakagawara, A. (1997) Interleukin-1 beta converting enzyme (ECE) is preferentially expressed in neuroblastomas with favourable prognosis. *Eur. J. Cancer* **33:** 2081–2083.

Ishikawa, E., Imagawa, M., Hashida, S., Yoshitake, S., Hamguchi, Y. and Ueno, T. (1983) Enzyme-labeling of antibodies and their fragments for enzyme immunoassay and immunohistochemical staining (review). *J. Immunoassay* **4:** 209–327.

Kerstens, H.M.J., Poddighe, P.J. and Hanselaar, A.G.J.M. (1995) A novel *in situ* hybridization signal amplification method based on the deposition of biotinylated tyramide. *J. Histochem. Cytochem.* **43:** 347–352.

Kiss, A., Palkovits, M. and Skirboll, L.R. (1988) Light microscopic triple-coloured immunohistochemical staining on the same vibratome section using the avidin–biotin–peroxidase complex technique. *Histochemistry* **88:** 353–356.

Kockx, M.M. (1998) Apoptosis in the atherosclerotic plaque: quantitative and qualitative aspects (review). *Art. Thromb. Vasc. Biol.* **18:** 1519–1522.

Köhler, G. and Milstein, C. (1975) Continuous cultures of fused cells secreting antibody of predefined specificity. *Nature* **256:** 495–497.

Krenács, T., Krenács, L., Bozóky, B. and Iványi, B. (1990) Double and triple immunocytochemical labelling at the light microscopical level in histopathology. *Histochem. J.* **22:** 530–536.

Kupper, H. and Storz, H. (1986) Double staining technique using a combination of indirect and direct immunofluorescence with monoclonal antibodies. *Acta. Histochem.* **78:** 185–188.

Lan, H.Y., Mu, W., Nikolic-Paterson, D.J. and Atkins, R.C. (1995) A novel, simple, reliable, and sensitive method for multiple immunoenzyme staining: use of microwave oven heating to block antibody crossreactivity and retrieve antigens. *J. Histochem. Cytochem.* **43:** 97–102.

Lanciego, J.L., Goede, P.H. and Wouterlood, F.G. (1997) Use of peroxidase substrate Vector VIP for multiple staining in light microscopy. *J. Neurosc. Meth.* **74:** 1–7.

Larison, K.D., BreMiller, R., Wells, K.S., Clements, I. and Haugland, R.P. (1995) Use of a new fluorogenic phosphatase substrate in immunohistochemical applications. *J. Histochem. Cytochem.* **43:** 77–93.

Larsson, L-I. (1988) *Immunocytochemistry: Theory and Practice*, pp. 171–179. CRC Press, Boca Raton, FL.

Larsson, L-I. (1993) Tissue preparation methods for light microscopic immunohisto-chemistry. *Applied Immunohistochem.* **1:** 1–16.

Lechago, J., Sun, N.C.Y. and Weinstein, W.M. (1979) Simultaneous visualization of two antigens in the same tissue section by combining immunoperoxidase with immuno-fluorescence techniques. *J. Histochem. Cytochem.* **27:** 1221–1225.

Lehr, H-A., Mankoff, D.A., Corwin, D., Santeusanio, G. and Gown, A.M. (1997) Application of Photoshop-based image analysis to quantification of hormone receptor expression in breast cancer. *J. Histochem. Cytochem.* **45:** 1559–1565.

Lehr, H-A., Van der Loos, C.M., Teeling, P. and Gown, A.M. (1999) Differential chromogen display and analysis in double immunohistochemical stains using Adobe Photoshop. *J. Histochem. Cytochem.* **47:** 119–125.

Leunissen, J.L.M. (1990) Background suppression using Aurion BSA-C and/or Tween 20. *Aurion Newsletter*, No. 1, Wageningen, The Netherlands.

Leunissen, J.L.M., Van de Plas, P.F.E.M. and Borghgraef, P.E.J. (1989) AuroProbe One, a new and universal ultra small gold particle-based (immuno)detection system for high sensitivity an improved penetration, in *Aurofile 2*, pp. 1–2. Janssen Life Sciences, Beerse, Belgium.

Li, C-Y., Ziesmer, S.C. and Lazcano-Villareal, O. (1987) Use of azide and hydrogen peroxide as inhibitor for endogenous peroxidase in the immunoperoxidase method. *J. Histochem. Cytochem.* **35:** 1457–1460.

Marijianowski, M.M.H., Teeling, M.M., Dingemans, K.P. and Becker, A.E. (1996) Multiple labeling in electron microscopy: its application in cardiovascular research. *Scanning Microsc. Suppl.* **10:** 261–271.

Marijianowski, M.M.H., Van Laar, M., Bras, J. and Becker, A.E. (1995) Chronic congestive heart failure is associated with a phenotypic shift of the intramyocardial endothelial cell. *Circulation* **92:** 1494–1498.

Malik, N.J. and Daymon, M.E. (1982) Improved double immunoenzyme labelling using alkaline phosphatase and horseradish peroxidase. *J. Clin. Pathol* **35:** 1092–1094.

Mason, D.Y., Abdulaziz, Z., Falini, B. and Stein, H. (1983) Double immunoenzymatic labelling, in *Immunocytochemistry: Practical Applications in Pathology and Biology* (J.M. Polak and S. Van Noorden eds), 1st edn, pp. 113. Wright, Bristol. UK.

Mason, D.Y. and Sammons, R. (1978) Alkaline phosphatase and peroxidase for double immunoenzymatic labelling of cellular constituents. *J. Clin. Pathol.* **31:** 454–460.

Mason, D.Y. and Woolston, R-E. (1982) Double immunoenzymatic labelling, in *Techniques in Immunohistochemistry* (G.R. Bullock and P. Petrusz eds), Vol. 1, pp. 135. Academic Press, London.

Mason, T.E., Phifer, R.F., Spicer, S.S., Swallow, R.S. and Dreskin, R.D. (1969) An immunoglobulin-enzyme bridge method for localizing tissue antigens. *J. Histochem. Cytochem.* **17:** 563–569.

McGadey, J. (1970) A tetrazolium method for non-specific alkaline phosphatase. *Histochemistry* **23:** 180–184.

Mesa-Tejada, R., Pascal, R.R. and Fenoglio, C.M. (1977) Immunoperoxidase: a sensitive immunohistochemical technique as a 'special stain' in the diagnostic pathology laboratory. *Hum. Pathol.* **8:** 313–320.

Mepham, B.L. (1982) Influence of fixatives on the immunoreactivity of paraffin sections. *Histochem. J.* **14:** 731–737.

Mepham, B.L., Frater, W. and Mitchell, B.S. (1979) The use of proteolytic enzymes to improve immunoglobulin staining by the PAP technique. *Histochem. J.* **11:** 345–357.

Miller, H.R.P. (1972) Fixation and tissue preservation for antibody studies. *Histochem. J.* **4:** 305–320.

Morgan, J.M., Navabi, H. and Jasani, A. (1997) Role of calcium chelation in high-temperature antigen retrieval at different pH values. *J. Pathol.* **182:** 233–237.

Mullink, H., Boorsma, D.M., Hensen-Logmans, S.C. and Meijer, C.J.L.M. (1987) Double immunoenzyme staining methods with special reference to monoclonal antibodies, in *Application of Monoclonal Antibodies in Tumor Pathology* (D.J. Ruiter, F.J. Fleuren and S.O. Warnaar eds), pp. 37–47. Martinus Nijhoff Publishers, Dordrecht, The Netherlands.

Naiem, M., Gerdes, J., Abdulaziz, Z., Sunderland, C.A., Stein, H. and Mason, D.Y. (1982) The value of immunohistological screening in the production of monoclonal antibodies. *J. Immunol. Meth.* **50:** 145–160.

Nakane, P.K. (1968) Simultaneous localization of multiple tissue antigens using the peroxidase-labeled antibody method: a study on pituitary glands of the rat. *J. Histochem. Cytochem.* **16:** 557–559.

Nakane, P.K. and Pierce, G.B. (1967) Enzyme-labeled antibodies: preparation and application for the localization of antigens. *J. Histochem. Cytochem.* **14:** 929–931.

Nemes, Z. (1987) Intensification of 3,3′-diaminobenzidin precipitation using the ferric ferricyanide reaction, and its application in the double-immunoperoxidase technique. *Histochemistry* **86:** 415–419.

Newman, G.R. and Jasani, B. (1984) Post-embedding immunoenzyme techniques, in *Immunolabeling for Electron Microscopy* (J.M. Polak and I.M. Varndell), pp. 53. Elsevier Scientific, Amsterdam.

Newman, G.R., Jasani, B. and Williams, E.D. (1989) Multiple hormone storage by cells of the human pituitary. *J. Histochem. Cytochem.* **37:** 1183–1192.

Norton, A.J. (1993) Microwave oven heating for antigen unmasking in routinely processed tissue sections (editorial). *J. Pathol.* **171:** 79–80.

Oliver, A.M. (1990) Macrophage heterogeneity in human fetal tissue. *Clin. Exp. Immunol.* **80:** 454–459.

Pastore, J.N., Clampett, C., Miller, J., Porter, K. and Miller, D. (1995) A rapid immuno-enzyme double labeling technique using EPOS reagents. *J. Histotechnol.* **27:** 1424–1429.

Polak, J.M. and Van Noorden, S. (1997) *Introduction to Immunocytochemistry*, 2nd Edn. Bios, Oxford, UK.

Polak, J.M. and Varndell, I.M. (1984) *Immunolabelling for Electron Microscopy*. Elsevier Scientific, Amsterdam.

Poletti, A., Manconi, R., Volpe, R., Sulfaro, S. and Carbone, A. (1987) Re: Application of double immunohistologic staining with monoclonal antibodies in lymph node pathology (letter to the editor). *J. Immunol. Meth.* **96:** 279–281.

Pryzwanski, K.B. (1982) Applications of double-label immunofluorescence, in *Techniques in Immunohistochemistry* (G.R. Bullock and P. Petrusz eds). Vol. 1, pp. 77–89. Academic Press, London.

Raap, A.K., Van de Corput, M.P.C., Vervenne, R.A.W., Van Gijlswijk, R.P.M., Tanke, H.J. and Wiegant, J. (1995) Ultra-sensitive FISH using peroxidase-mediated deposition of biotin- or fluorochrome tyramide. *Hum. Mol. Genet.* **4:** 529–534.

Richter, S. and Mir, R. (1984) A double-staining peroxidase-antiperoxidase technique. *J. Histotechnol.* **7:** 8–9.

Robben, H., Van Dekken, H., Poddighe, P.J. and Vooijs, G.P. (1994) Identification of aneuploid cells in cytological specimens by combined *in situ* hybridization and immunocytochemistry. *Cytopathology* **5:** 384–391.

Rye, D., Saper, C.B. and Wainer, B.H. (1984) Stabilization of tetramethylbenzidine (TMB) reaction product: application for retrograde and anterograde tracing, and combination with immunohistochemistry. *J. Histochem. Cytochem.* **32:** 1145–1153.

Sabattini, E., Bisgaard, K., Ascani, S., Poggi, S., Piccioli, M., Ceccarelli, C., Pieri, F., Fraternali-Ocioni, G. and Pileri, S.A. (1998) The EnVision™+ system: a new immunohistochemical method for diagnostics and research. Critical comparison with the APAAP, ChemMate™, CSA, LABC, and SABC techniques. *J. Clin. Pathol.* **51:** 506–511.

Sakanaka, M., Magari, S., Shibasaki, T., Shinoda, K. and Kohno, J. (1988) A reliable method combining horseradish peroxidase histochemistry with immuno-β-galactosidase staining. *J. Histochem. Cytochem.* **36:** 1091–1096.

Scopsi, L. and Larsson, L-I. (1986) Increased sensitivity in peroxidase immunocytochemistry. A comparative study of a number of peroxidase visualization methods. *Histochemistry* **84:** 221–230.

Shi, S-R., Cote, R.J. and Taylor, C.R. (1997) Antigen retrieval immunohistochemistry: past, present and future. *J. Histochem. Cytochem.* **45:** 327–343.

Shi S-R., Imam, A., Young, L., Cote, R.J. and Taylor, C.R. (1995) Antigen retrieval immunohistochemistry under the influence of pH using monoclonal antibodies. *J. Histochem. Cytochem.* **43**: 193–201.

Shi, S-R., Itzkowitz, S.H. and Kim, Y.S. (1988) A comparison of three immunoperoxidase techniques for antigen detection in colorectal carcinoma tissues. *J. Histochem. Cytochem.* **36**: 317–322.

Shi, S-R., Key, M.E. and Kalra, K.L. (1991) Antigen retrieval in formalin-fixed, paraffin-embedded tissues: an enhancement method for immunohistochemical staining based on microwave oven heating of tissue sections. *J. Histochem. Cytochem.* **39**: 741–748.

Silverstein, A.M. (1957) Contrasting fluorescent labels for two antibodies (letter). *J. Histochem. Cytochem.* **5**: 94–95.

Sofroniew, M.V. and Schrell, U. (1982) Long-term storage and regular repeated use of diluted antisera in glass staining jars for increased sensitivity, reproducibility, and convenience of single- and two-color light microscopic immunocytochemistry. *J. Histochem. Cytochem.* **30**: 504–511.

Speel, E-J., Schutte, B., Wiegant, J., Raemakers, F.C.S. and Hopman, A.H.N. (1992) A novel fluorescence detection method for *in situ* hybridization, based on the alkaline phosphatase-Fast Red reaction. *J. Histochem. Cytochem.* **40**: 1299–1308.

Speel, E.J.M., Raemakers, F.C.S. and Hopman, A.H.N. (1995) Cytochemical detection systems for *in situ* hybridization, and the combination with immunocytochemistry. 'Who is afraid of red, green and blue?' *Histochem. J.* **27**: 833–858.

Sternberger, L.A., Hardy, P.H., Cuculis, J.J. and Meyer, H.G. (1970) The unlabeled antibody enzyme method of immunohistochemistry. Preparation and properties of soluble antigen-antibody complex (horseradish peroxidase-anti-horseradish peroxidase) and its use in identification of spirochetes. *J. Histochem. Cytochem.* **18**: 315–333.

Sternberger, L.A. and Joseph, S.A. (1979) The unlabeled antibody method. Contrasting colour staining of paired pituitary hormones without antibody removal. *J. Histochem. Cytochem.* **27**: 1424–1429.

Sträter, J., Günthert, A.R. and Brüderlein, S. (1995) Microwave irradiation of paraffin-embedded tissue sensitizes the TUNEL method for *in situ* detection of apoptotic cells. *Histochemistry* **103**: 157–160.

Streefkerk, J. (1972) Inhibition of erythrocyte pseudoperoxidase activity by treatment with hydrogen peroxide following methanol. *J. Histochem. Cytochem.* **20**: 829–831.

Suurmeijer, A.J.H. and Boon, M.E. (1993) Optimizing keratin and vimentin retrieval in formalin-fixed, paraffin-embedded tissue with the use of heat and metal salts. *Appl. Immunohistochem.* **1**: 143–148.

Taylor, C.R. (1974) The nature of Reed-Sternberg cells and other malignant 'reticulum' cells. *Lancet* **2**: 802–807.

Taylor, C.R. (1978) Immunoperoxidase techniques. Practical and theoretical aspects. *Arch. Pathol. Lab. Med.* **102**: 113–121.

Taylor, C.R. (1980) Immunohistological studies for lymphoma: past, present and future. *J. Histochem. Cytochem.* **28**: 777–787.

Taylor, C.R. and Burns, J. (1974) The demonstration of plasma cells and other immunoglobulin-containing cells in formalin-fixed, paraffin-embedded tissues using peroxidase labelled antibody. *J. Clin. Pathol.* **27**: 14–20.

Taylor, C.R. and Mason, D.Y. (1974) The immunohistological detection of intracellular immunoglobulin in formalin-paraffin sections from multiple myoloma and related conditions using the immunoperoxidase technique. *Clin. Exp. Immunol.* **18**: 417–429.

Tharp, M.D., Seelig, L.L., Tigelaar, R.E. and Bergstresser, P.R. (1985) Conjugated avidin binds to mast cell granules. *J. Histochem. Cytochem.* **33**: 27–32.

The, T.H. and Feldkamp, T.E.W. (1970) Conjugation of fluorescein isothiocyanate to antibodies. *Immunology* **18**: 875–881.

Thisted, M. (1995) The use of levamizole in immunohistochemistry. *Clin. Lab. Int.* **19**: 37.

Tidman, N., Janossy, G., Bodger, M., Kung, P.C. and Goldstein, G. (1981) Delineation of human thymocyte differentiation pathways utilizing double staining techniques with monoclonal antibodies. *Clin. Exp. Immunol.* **45**: 457–467.

Tramu, G., Pillez, A. and Leonardelli, J. (1978) An efficient method of antibody elution for the successive or simultaneous localization of two antigens by immunocytochemistry. *J. Histochem. Cytochem.* **26:** 322–324.

Trojanowski, J.Q., Obrocka, M.A. and Lee, V.M. (1983) A comparison of eight different chromogen protocols for the demonstration of immunoreactive neurofilaments or glial filaments in rat cerebellum using the peroxidase-antiperoxidase method and monoclonal antibodies. *J. Histochem. Cytochem.* **31:** 1217–1223.

Vacca, L.L., Abrahams, S.J. and Naftchi, N.E. (1980) A modified peroxidase-antiperoxidase procedure for improved localization of tissue antigens. *J. Histochem. Cytochem.* **28:** 297–307.

Valnes, K. and Brandtzaeg, P. (1982) Comparison of paired immunofluorescence and paired immunoenzyme staining methods based on primary antisera from the same species. *J. Histochem. Cytochem.* **30:** 518–524.

Valnes, K. and Brandtzaeg, P. (1984) Paired indirect immunoenzyme staining with primary antibodies from the same species. Application of horseradish peroxidase and alkaline phosphatase as sequential labels. *Histochem. J.* **16:** 477–487.

Valnes, K., Brandtzaeg, P. and Rognum, T.O. (1984) Sensitivity and efficiency of four immunohistochemical methods as defined by staining of artificial sections. *Histochemistry* **81:** 313–319.

Van den Brink, G.R., Tytgat, K.M.A.J., Van der Hulst, R., Van der Loos, C.M., Einerhand, A.W.C., Büller, H.A. and Dekker, J. (1999) Localization of two human gastric mucins, MUC5AC and MUC6, and *Helicobacter pylori* reveals exclusive co-localization of *H. pylori* with MUC5AC (abstract). *Eur. J. Gastro. Hepatol.* **10:** A78–A79.

Van den Brink, W.J., Van der Loos, C.M., Volkers, H.H., Lauwen, R., Van den Berg, F.M., Houthoff, H-J. and Das, P.K. (1990) Combined β-galactosidase and immunogold/silver staining for immunohistochemistry and DNA *in situ* hybridization. *J. Histochem. Cytochem.* **38:** 325–329.

Van den Brink, G.R., Tytgat, K.M.A.J., Van der Hulst, R., Van der Loos, C.M., Einerhand, A.W.C., Büller, H.A. and Dekker, J. *Helicobacter pylori* co-localizes with MUC5AC in the human stomach, *Gut* (Submitted).

Van den Oord, J.J., Faccetti, F., De Wolf-Peeters, C. and Desmet, V.J. (1989) Binding of biotin to hepatitis B surface antigen: a possible pitfall in immunohistochemistry. *J. Histochem. Cytochem.* **37:** 551–554.

Van de Plas, P.F.E.M. and Leunissen, J.L.M. (1989) Immunocytochemical detection of tubilin in whole mount preparation of PtK2-cells: improved penetration characteristics of AuroProbe One. *Aurofile 2*, pp. 3–4. Janssen Life Sciences, Beerse, Belgium.

Van der Loos, C.M. (1992) *Development of Novel Immunohistochemical and Hybrido-cytochemical Multiple Staining Methods (PhD thesis).* Free University Press, Amsterdam.

Van der Loos, C.M. (1998) *Immunoenzymatic Double Staining Methods. A Practical Guide.* DAKO Handbook 00105, DAKO A/S, Glostrup, Denmark.

Van der Loos, C.M. and Becker, A.E. (1994) Double epi-illumination microscopy with separate visualization of two antigens: a combination of epi-polarization for immunogold-silver staining and epi-fluorescence for alkaline phosphatase staining. *J. Histochem. Cytochem.* **42:** 289–295.

Van der Loos, C.M., Becker, A.E. and Van den Oord, J.J. (1993) Practical suggestions for successful immunoenzyme double staining experiments. *Histochem. J.* **25:** 1–13.

Van der Loos, C.M., Das, P.K. and Houthoff, H-J. (1987) An immunoenzyme triple staining method using both polyclonal and monoclonal antibodies from the same species. Application of combined direct, indirect and avidin-biotin complex (ABC) technique. *J. Histochem. Cytochem.* **35:** 1199–1204.

Van der Loos, C.M., Das, P.K., Van den Oord, J.J. and Houthoff, H-J. (1989a) Multiple immunoenzyme staining techniques. Using of fluoresceinated, biotinylated and unlabeled monoclonal antibodies. *J. Immunol. Meth.* **117:** 45–52.

Van der Loos, C.M., Naruko, T. and Becker, A.E. (1996) The use of enhanced polymer one-step staining (EPOS) reagents for immunoenzyme double-labelling. *Histochem. J.* **28:** 709–714.

Van der Loos, C.M., Van den Oord, J.J., Das, P.K. and Houthoff, H-J. (1988) Use of commercially available monoclonal antibodies for immunoenzyme double staining. *Histochem. J.* **20:** 409–413.

Van der Loos, C.M., Volkers, H.H., Rook, R., Van den Berg, F.M. and Houthoff, H-J. (1989b) Simultaneous application of DNA *in situ* hybridization and immunohistochemistry on one tissue section. *Histochem. J.* **21:** 279–284.

Vandesande, F. (1983) Immunohistochemical double staining techniques, in *Immunohisochemistry. IBRO Handbook* (A.C. Cuello ed), pp. 257–272. John Wiley & Sons, Chichester, UK.

Van Noorden, S. (1986) Tissue preparation and immunostaining techniques for light microscopy, in *Immunocytochemistry. Modern Methods and Applications* (J.M. Polak and S. Van Noorden eds), 2nd edn, pp. 26–53. Wright, Bristol, UK.

Van Noorden, S., Stuart, M.C., Cheung, A., Adams, E.F. and Polak, J.M. (1986) Localization of pituitary hormones by multiple immunoenzyme staining procedures using monoclonal and polyclonal antibodies. *J. Histochem. Cytochem.* **34:** 287–292.

Vyberg, M. and Nielsen, S. (1998) Dextran polymer conjugate two-step visualization system for immunohistochemistry. *Appl. Immunohistochem.* **6:** 3–10.

Wallace, E. and Wofsy, L. (1979) Hapten-sandwich labeling. IV. Improved procedures and non-cross-reacting hapten reagents for double-labeling cell surface antigens. *J. Immunol. Meth.* **25:** 283–289.

Warnke, R. and Levy, R. (1980) Detection of T and B cell antigens with hybridoma monoclonal antibodies: a biotin-avidin-horseradish peroxidase method. *J. Histochem. Cytochem.* **28:** 771–776.

Werner, M, Von Wasielewski, R. and Komminoth, P. (1996) Antigen retrieval, signal amplification and intensification in immunohistochemistry. *Histochem. Cell. Biol.* **105:** 253–260.

Wofsy, L., Henry, C. and Cammisuli, S. (1978) Hapten-sandwich labeling of cell-surface antigens. *Contemp. Top. Mol. Immunol.* **7:** 215–237.

Wolber, R.A., Lloyd, R.V. (1988) Cytomegalovirus detection by non-isotopic DNA *in situ* hybridization and viral antigen immunostaining using a two-colour technique. *Hum. Pathol.* **19:** 736–741.

Wood, G.S. and Warnke, R. (1981) Suppression of endogenous avidin-binding activity in tissues and its relevance to biotin–avidin detection systems. *J. Histochem. Cytochem.* **29:** 1196–1204

Appendix B

B.1. Buffers/fixatives

B.1.1 Tris-buffered saline (TBS)

$10 \times$ concentrated.
  Dissolve 303 g Tris and 450 g NaCl in $\approx$ 4 l distilled water.
  Adjust pH to 7.8 with $\approx$ 185 ml HCl (32%).
  Bring volume to 5 l with distilled water.

  Dilute $10 \times$ with distilled water before use.

B.1.2 Phosphate-buffered saline (PBS)

$10 \times$ concentrated.
  Dissolve 70.5 g disodium hydrogen phosphate dihydrate, 10.5 g potassium dihydrogen phosphate and 450 g NaCl in $\approx$ 4 l distilled water.
  Adjust pH to 7.4.
  Bring volume to 5 l with distilled water.

  Dilute $10 \times$ with distilled water before use. Check pH to be 7.4.

B.1.3 Citrate buffer for heat-induced antigen retrieval

Dissolve 2.1 g citric acid monohydrate (100 mM) in 1 l distilled water.
  Adjust pH to 6.0 with $\approx$ 13 ml 2 N NaOH solution.

B.1.4 TdT-buffer (pH 7.2) for TUNEL

Dissolve in 50 ml distilled water.
  5.35 g sodium cacodylate.$3H_2O$ (500 mM); 11.5 mg Cobalt (II) chloride (1 mM); 125 µl BSA 20% (0.05%); and 0.44 g NaCl (150 mM).

B.1.5 TB-buffer for TUNEL

Dissolve in 50 ml distilled water.
0.88 g NaCl (300 mM); and 0.44 g sodium citrate dihydrate (30 mM).

B.1.6 Zamboni fixative

NaOH	4.3 g
Distilled water	800 ml
Paraformaldehyde	20 g
Sodium dihydrogen phosphate dihydrate	18.8 g
Saturated picric acid solution (double filtered)	150 ml

Dissolve NaOH in distilled water and then add paraformaldehyde. Stir until completely dissolved. Add sodium dihydrogen phosphate dihydrate and stir until dissolved. Add picric acid solution, check and/or adjust pH to 7.3 and bring volume to 1 l with distilled water.

The solution is stable for 1 year at room temperature.

B.1.7 Methacarn fixative

Methanol (*)	600 ml
Chloroform	300 ml
Acetic acid (glacial)	100 ml

The solution is stable for 1 year at room temperature.

(*) Methanol can be replaced by regular ethylalcohol 96%; the fixative is then named 'Carnoy'. Any differences between methacarn and Carnoy are unknown.

B.1.8 Methylgreen nuclear counterstain

Methylgreen, Crystal Violet-free (Sigma M6776).
0.1% Methylgreen in sodium acetate buffer (100 mM, pH 5.2).
This solution is stable at 4°C for at least 1 month.

Cover sections with sodium acetate buffer (10 mM, pH 5.2)	15 min
Blott off, do not wash	
Cover sections with Methylgreen solution	5 min (*)
Wash briefly with distilled water	
Preferably, dehydrate and mount organically (**)	

(*): Paraffin sections which have been pretreated with heat-induced antigen retrieval using citrate pH 6.0, will show a weak nuclear counter stain. Prolong the staining time up to 10–15 min.

(**): When using this nuclear counterstain in combination with alcohol-soluble chromogens, the tissue specimen should be post-fixed with buffered formalin, rinsed with distilled water and aqueously mounted. Nuclear counterstaining intensity will decrease in time, but will not fade away completely.

Appendix C

C.1 General immunohistochemistry

C.1.1. Cryostat sections

1. Mount 4–6 µm cryostat sections on coated* microscopic slide.
2. Air dry for at least 1–2 h, preferably overnight under a fan.
3. Wrap in Parafilm–foil 'face-to-face' (use small carton strips at the glass edges) and store dry in a closed box at $-80°C$ or go to step 5.
4. Make sure that Parafilm–foil-packed tissue specimens reach room temperature before unwrapping.
5. Fix in acetone (*p.a.* quality) for 10 min at 4°C.
6. Air-dry again briefly (2 min).
7. Encircle the tissue sections with a water-repellant rim (e.g. commercially available from DAKO).
   Option:
   Apply extra fixation with Zamboni fixative (1–2 min at room temperature) (see *Appendix B.1.6*) and rinse with washing buffer (3×2 min).
8. Block endogenous peroxidase activity using sodium azide (0.1%) + hydrogen peroxide (0.3%) in washing buffer for 20 min at room temperature (Li *et al.*, 1987).
9. Rinse with washing buffer (3×2 min).
10. Carry out the immunohistochemical procedure including enzyme activity detection.
    Option:
    Post-fix with 4% buffered formalin for 5 min.
11. Wash with running tap water for 5 min.
12. Coverslip with an appropriate mounting medium.

C.1.2 Paraffin sections

1. Mount 4–6 µm paraffin sections on coated* microscope slide.
2. Dry for 1 h at 60°C or overnight at 37°C.
3. Deparaffinize and rehydrate section.
4. Block endogenous peroxidase activity using 0.3% hydrogen peroxide in methanol for 20 min at room temperature (Streefkerk, 1972). Rinse with running tap water.
5. Treatment with pepsin, trypsin or heat-induced antigen retrieval (see below), depending on the antigen to be detected. These treatments are not useful for methacarn-fixed tissue blocks.
6. Rinse with running tap water.
7. Dry carefully around tissue section and encircle with water-repellant rim.
8. Rinse with washing buffer.
9. Carry out the immunohistochemical procedure including enzyme activity detection.
10. Wash with running tap water for 5 min.
11. Coverslip with an appropriate mounting medium.

Pepsin pretreatment. Dissolve 0.25% pepsin (e.g. Sigma P7000) in 10 mM HCl. Incubate tissue sections for 15 min at 37°C.

Trypsin pretreatment. Dissolve 0.1% trypsin (e.g. Sigma T8128) in 50 mM Tris–HCl buffer pH 7.8. Incubate tissue sections for 10 min at 37°C.

Heat-induced antigen retrieval. Use citrate buffer pH 6.0 (*Appendix B.1.3*) or 10 mM Tris/1 mM EDTA pH 9.0 (Vyberg and Nielsen, 1998). Heat citrate buffer and sections to 100°C in household microwave oven at maximum power. Upon boiling, turn down power and keep the solution boiling for 15 min. Remove sections from citrate buffer after 10 min cool-down.

*Coating either with: organosilan, poly-L-lysine, or commercially coated microscope slides.

Appendix D

D.1. Double-staining protocols

General comment on the double-staining protocols Appendix D.1.1.–D.1.5. Primary antibodies are incubated for either 30–60 min at room temperature or overnight at 4°C in a humidified chamber. Room temperature should be applied if no temperature is indicated.

D.1.1 Sequential immunoenzyme double staining

The sequential double-staining concept is based on two individually subsequently performed complete immunostaining procedures, with or without an antibody elution step in between. The procedure is fully independent of the primary antibody species, Ig type or IgG isotype (Nakane, 1968; Sternberger and Joseph, 1979).

Good results have been obtained using the newly designed DAKO Doublestain System based on EnVision^{TM} technology (DAKO code no. K1395):

1. Non-specific binding blocking step, tap off do not wash   15 min
2. Mouse or rabbit antibody 1   overnight 4°C or 30–60 min
   wash with TBS   3 × 2 min
3. EnVision^{TM}/HRP, GAM+GAR   30 min
   wash with TBS   3 × 2 min
4. Development of HRP activity with DAB in brown   2–10 min
5. Elution step with DAKO double staining block   30 min
   alternatively, 5 min boiling in citrate pH 6.0.
   (*Appendix B.1.3* and *C.1.2* and Lan *et al.*, 1995.)
   wash with distilled water   5 min
6. Non-specific binding blocking step, tap off do not wash   5 min
7. Mouse or rabbit antibody 2   4°C or 30–60 min
   wash with TBS   3 × 2 min
8. EnVision^{TM}/AP, GAM/GAR   60 min
   wash with TBS   3 × 2 min
9. Development of AP activity detection with Fast Red in red*   5–30 min

*Instead of Fast Red, New Fuchsin, Fast Blue BB or NBT/BCIP can be used (*Table 5.5a*).

(See *Figure 3.1* for a schematic drawing of this procedure.)

Comment. The commercially available kit is a user-friendly system for the evaluation of two different cell populations, but is less recommended for those instances where mixed coloured products are expected at sites of co-localization.

D.1.2 Simultaneous immunoenzyme double staining

Direct/direct concept based on: two directly labelled primary monoclonal antibodies (Boorsma, 1984).

1. Non-specific binding blocking step, tap off do not wash   15 min
2. MAb 1/conjugate 1 + MAb 2/conjugate 2   overnight 4°C or 60 min
   ('cocktail')
   wash with TBS   3 × 2 min
   further steps are primary antibody-label dependent.
3. First enzyme activity detection,
   wash with TBS   3 × 2 min
4. Second enzyme activity detection or silver enhancement.

Comment. The markers involved with directly labelled primary antibodies can be either enzymes, biotin (with a streptavidin reagent as second layer), or fluorochromes/haptens (with anti-fluorochrome or anti-hapten as second layer). DAKO EPOS™/HRP reagents can be regarded as directly HRP-conjugated primary antibodies (Pastore *et al.*, 1995; Van der Loos *et al.*, 1997), and can therefore be incorporated into this type of protocol.

Example with two fluorochromes (see *Plate 5*)

1. Normal swine serum, tap off do not wash   15 min
2. MAb1/FITC-conjugate + MAb2/Phycoerythrin-conjugate   60 min
   wash with TBS   3 × 2 min
3. Goat anti-FITC + rabbit anti-Phycoerythrin   15 min
   wash with TBS   3 × 2 min
4. SAG/HRP (Biosource, Nivelles, Belgium) +
   SAR/AP (DAKO)   30 min
   wash with TBS   3 × 2 min
5. AP activity detection in blue (Fast Blue)
   wash with TBS   3 × 2 min
6. HRP activity detection in red (AEC)

(See *Figure 3.2* for schematic drawing.)

Reagents	Code no.	Dilution
Streptavidin/HRP	DAKO P0397	1:400
Streptavidin/AP	DAKO D0396	1:100
Streptavidin–biotin complex/HRP	DAKO K0377	1:100
Streptavidin/GAL	Roche/Boehringer 1 112 481	1:40
Goat anti-biotin/ultra-small gold*	Aurion 800.088	1:40
Rabbit anti-biotin/HRP	DAKO P5106	1:50
Rabbit anti-biotin/AP	DAKO D5107	1:20
Sheep anti-DIG/HRP	Roche/Boehringer 1 207 733	1:400
Sheep anti-DIG/AP	Roche/Boehringer 1 093 274	1:100
Sheep anti-DIG/ultra-small gold	Aurion 800.199	1:40
Rabbit anti-FITC	DAKO V0403	1:1000
Rabbit anti-FITC/HRP	DAKO P5100	1:50
Rabbit anti-FITC/AP	DAKO D5101	1:20
Goat anti-FITC	BioGenesis	1:400
Mouse anti-FITC/ultra-small gold	Aurion 800.233	1:40
Rabbit anti-Phycoerythrin	BioGenesis	1:200
for more anti-fluorochrome antibodies	see also BioGenesis	
Swine anti-goat Ig/HRP	Biosource ACI4304	1:50
Swine anti-goat Ig/AP	Biosource ACI3405	1:10
Swine anti-rabbit Ig/HRP	DAKO P0399	1:50
Swine anti-rabbit Ig/AP	DAKO D0306	1:10

*It is proven empirically that the use of anti-biotin/USG is far more efficient than streptavidin/USG.

D.1.3 Simultaneous immunoenzyme double staining

Indirect/indirect concept based on two unlabelled primary antibodies of different animal host origin (Campbell and Bhatnagar, 1976; Mason and Sammons, 1978).

1. Non-specific binding blocking step with normal goat serum $\qquad$ 15 min
   tap off, do not wash
2. Mouse antibody 1 + rabbit antibody 2 $\qquad$ overnight 4°C or 60 min
   ('cocktail')
   wash with TBS $\qquad$ 3×2 min
3. GAM-Enzyme I conjugated + GAR-Enzyme II conjugated $\qquad$ 30 min
   ('cocktail')
   wash with TBS $\qquad$ 3×2 min
4. First enzyme activity detection
   wash with TBS $\qquad$ 3×2 min
5. Second enzyme activity detection or silver enhancement.

(See *Figures 3.3* and *3.4* for schematic drawing.)

Comment. To prevent cross-reaction between second step reagents from different species, it is recommended that reagents are applied that have been raised in the same host.

Reagents	Code no.	Dilution
Normal goat Serum	DAKO X0907	1:10
Goat anti-mouse Ig/HRP	DAKO P0447	1:50
EnVision™+/HRP, goat anti-mouse Ig	DAKO K4000/4001	undiluted
Goat anti-mouse Ig/AP	DAKO D0486	1:20
Goat anti-mouse Ig/GAL	SBA 1010-06	1:10
Goat anti-mouse IgG + IgM/ultra-small gold	Aurion 800.044	1:40
Goat anti-rabbit Ig/HRP	DAKO P0448	1:50
EnVision™+/HRP, goat anti-rabbit Ig	DAKO K4002/4003	undiluted
Goat anti-rabbit Ig/AP	DAKO D0487	1:20
Goat anti-rabbit Ig/GAL	SBA 4010-06	1:10
Goat anti-rabbit Ig/ultra-small gold	Aurion 800.011	1:40

For more sensitivity, one biotinylated second step reagent (step 3) can be incorporated into this protocol. This step is followed by an approp-riate third step streptavidin- or anti-biotin reagent (*Appendix D.1.2*).

Reagents	Code no.	Dilution
Goat anti-mouse Ig/biotin	DAKO E0433	1:200
Goat anti-rabbit Ig/biotin	DAKO E0432	1:400

Alternatively, for more sensitivity, mouse APAAP and rabbit PAP comp-lexes can be applied (Malik and Daymon, 1982; Mason and Sammons, 1978).

Reagents	Code no.	Dilution
Goat anti-mouse Ig	DAKO Z0420	1:25
Goat anti-rabbit Ig	DAKO Z0421	1:25
APAAP, mouse monoclonal	DAKO D0651	1:50
PAP, rabbit polyclonal	DAKO Z0113	1:100

D.1.4 Simultaneous immunoenzyme double staining

Indirect/indirect concept based on two unlabelled primary mouse monoclonal antibodies of different Ig subtypes or IgG isotypes (Tidman *et al.*, 1981; Van der Loos *et al.*, 1993).

1. Non-specific binding blocking step with normal goat serum   15 min
   tap off, do not wash.
2. Mouse antibody 1 (Ig isotype or IgG subclass X) +   overnight at
   mouse antibody 2 (Ig isotype or IgG subclass Y)   4°C or 60 min
   ('cocktail')
   wash with TBS   3 × 2 min
3. GAM-IgX Enzyme I conjugated + GAM-IgY-Enzyme II
   conjugated ('cocktail')   30 min
   wash with TBS   3 × 2 min
4. First enzyme activity detection
   wash with TBS   3 × 2 min
5. Second enzyme activity detection or silver enhancement.

(See *Figure 3.5* for schematic drawing).

Comment. As with the previous *indirect/indirect* concept, more sensitivity for at least one primary antibody can be obtained with a biotinylated second step reagent (step 3). This step should be followed by an appropriate third step streptavidin- or anti-biotin reagent (*Appendix D.1.2*).

Reagents	Code no.	Dilution
Normal goat serum	DAKO X0907	1:10
Goat anti-mouse IgM/HRP	SBA 1020-05	1:50
Goat anti-mouse IgM/biotin	SBA 1020-08	1:100
Goat anti-mouse IgM/AP	SBA 1020-04	1:20
Goat anti-mouse IgG$_1$/HRP	SBA 1070-05	1:50
Goat anti-mouse IgG$_1$/biotin	SBA 1070-08	1:100
Goat anti-mouse IgG$_1$/AP	SBA 1070-04	1:20
Goat anti-mouse IgG$_{2a}$/HRP	SBA 1080-05	1:50
Goat anti-mouse IgG$_{2a}$/biotin	SBA 1080-08	1:100
Goat anti-mouse IgG$_{2a}$/AP	SBA 1080-04	1:20
Goat anti-mouse IgG$_{2b}$/HRP	SBA 1090-05	1:50
Goat anti-mouse IgG$_{2b}$/biotin	SBA 1090-08	1:100
Goat anti-mouse IgG$_{2b}$/AP	SBA 1090-04	1:20
Goat anti-mouse IgG$_3$/biotin	SBA 1100-08	1:100
Goat anti-mouse IgG$_3$/AP	SBA 1100-04	1:20

D.1.5 Multi-step immunoenzyme double staining

Indirect/direct multi-step concept based on one unlabelled monoclonal antibody and one FITC conjugated (Van der Loos *et al.*, 1989a).

1. Non-specific binding blocking step with normal goat serum   15 min
   tap off, do not wash.
2. Mouse MAb 1   overnight at 4˚C or 60 min
   wash with TBS 3×2 min
3. GAM/biotin   30 min
   wash with TBS 3×2 min
4. Normal mouse serum + streptavidin-Enzyme I reagent,
   (*Appendix D.1.2*)   30 min
   wash with TBS 3×2 min
5. Mouse MAb 2/FITC conjugated   overnight at 4˚C or 60 min
   wash with TBS 3×2 min
6. Rabbit anti-FITC   15 min
   wash with TBS 3×2 min
7. GAR-Enzyme II conjugated   30 min
   wash with TBS 3×2 min
8. First enzyme activity detection,
   wash with TBS 3×2 min
9. Second enzyme activity detection or silver enhancement.

(See *Figure 3.6* for schematic drawing)

Alternatives:
Three-step AP detection of MAb 1, avoiding streptavidin–biotin interaction:
3. GAM   30 min
   wash with TBS 3×2 min
4. APAAP complex, mouse   30 min
   wash with TBS 3×2 min
5. Normal mouse serum   15 min
   continue the above protocol with mouse MAb 2/FITC conjugated.

(See *Figure 3.7* for schematic drawing)

Two-step HRP detection of MAb 1, avoiding streptavidin–biotin interaction:
3. EnVision™+/HRP, GAM   30 min
   wash with TBS 3×2 min
4. Normal mouse serum   15 min
   continue the above protocol with mouse MAb 2/FITC conjugated.

(See *Figure 3.8* for schematic drawing)

Comments. Apart from directly FITC-conjugated primary antibodies, as frequently used for the Fluorescein Activated Cell Sorter, Texas Red-, Rhodamine-, AMCA-, or phycoerythrin-conjugated antibodies can also be applied for this double-staining technique by inserting an appropriate step 6 reagent. Biotinylated primary antibodies can also be used in step 5 (Van der Loos *et al.*, 1988), followed by a streptavidin- or anti-biotin reagent (*Appendix D.1.2*).

Like the DAKO EnVision™ products, the Enhanced Polymer One-Step (EPOS™) staining products basically consist of a dextran polymer backbone. With EPOS™ products (DAKO U-class reagents), this backbone is labelled with a variety of primary antibodies and HRP enzymes. EPOS™ products can be regarded as directly labelled primary antibodies, and can therefore be applied in step 5 of this double staining concept, without any further detection steps needed (Van der Loos *et al.*, 1996). See *Figure 3.9* for schematic drawing.

Reagents	Code no.	Dilution
Normal goat serum	DAKO X0907	1:10
Goat anti-mouse Ig/biotin	DAKO E0433	1:200
Goat anti-mouse Ig/HRP	DAKO P0447	1:50
Goat anti-mouse Ig/AP	DAKO D0486	1:20
Goat anti-mouse Ig/GAL	SBA 1010-06	1:10
Goat anti-mouse Ig	DAKO Z0420	1:25
Goat anti-mouse IgG + IgM/ultra-small gold	Aurion 800.044	1:40
Alkaline phosphatase anti-alkaline phosphatase, mouse	DAKO D0651	1:100
EnVision^{TM} + /HRP, goat anti-mouse Ig	DAKO K4000	undiluted
Goat anti-rabbit Ig/HRP	DAKO P0448	1:50
Goat anti-rabbit Ig/AP	DAKO D0487	1:20
Goat anti-rabbit Ig/GAL	SBA 4010-06	1:10
Goat anti-rabbit Ig/ultra-small gold	Aurion 800.011	1:40
Normal mouse serum	DAKO X0910	1:10
Rabbit anti-FITC	DAKO V0403	1:1000
Mouse anti-FITC/ultra-small gold	Aurion 800.233	1:40
Rabbit anti-FITC/HRP	DAKO P5100	1:50
Rabbit anti-FITC/AP	DAKO D5101	1:20

D.1.6 Special remarks for the application of ultra-small gold-conjugated reagents

When employing ultra-small gold-conjugated antibodies in single- or double-staining protocols, there should be special care taken to prevent non-specific binding, as a consequence of the negatively charged gold particles (Leunissen, 1990).

This protocol starts just before the application of the ultra-small gold conjugate, and can be inserted directly into single- and double-staining procedures. Because of the high sensitivity/efficiency of the IGS staining, it is recommended that longer washing steps (3×5 min instead of 3×2 min) than usual are used throughout the whole staining procedure, especially when epi-polarization microscopy will be employed for observation.

- IGS-blocking solution*　　20 min
- wash with IGS Buffer†　　3×5 min
- ultra-small gold-conjugated reagent, diluted in IGS Buffer　　120–150 min
- wash with IGS buffer　　4×5 min
- rinse with TBS
- post-fixation with buffered formalin　　10 min
- running tap water　　5 min
- rinse with distilled water
- silver enhancement with Aurion R-Gent　　5–20 min
  mix just before use equal amounts of developer and enhancer
  check staining intensity microscopically
- wash with running tap water　　10 min
  (repeat silver enhancement with fresh reagent if staining is
  still weak after 20 min).

*IGS blocking: TBS + 0.8% CWFSGEL + 5% BSA + NGS (1:20) + 0.1% Na-azide.
†IGS buffer: TBS (pH 7.2–7.4) + 0.1% BSA-C + 0.1% Na-azide.

Reagents	Code no.	Dilution
Goat anti-rabbit Ig/ultra-small gold	Aurion 100.011	1:50
Goat anti-mouse Ig/ultra-small gold	Aurion 100.044	1:50
Rabbit anti-goat Ig/ultra-small gold	Aurion 100.077	1:50
Goat anti-biotin/ultra-small gold		
(for three-step detection via biotin)	Aurion 100.088	1:50
Sheep anti-DIG/ultra-small gold	Aurion 100.199	1:50
R-GENT silver-enhancement kit	Aurion 500.011	
BSA-C (acetylated BSA)	Aurion 900.099	
CWFSGel (cold water fish skin gelatin)	Aurion 900.033	

D.1.7 Combination of TUNEL and immunohistochemistry

Concept based on first TUNEL, without detection, followed by immuno-histochemistry, and ending with enzymatic visualizations (Van der Loos and Houtkamp, unpublished). The TUNEL protocol is according to Gavrieli *et al*. (1992), using Roche/Boehringer TdT kit no. 220 582.

1. Deparaffinize, and rehydrate tissue section.
2. Pretreatment with Protease K (20 μg ml^{-1} in PBS), dependent on the tissue type — 5–20 min, 22°C rinse with running tap water.
3. Block endogenous peroxidase activity using 0.3% hydrogen peroxide in methanol — 20 min rinse with running tap water.
4. Dry carefully around tissue section and encircle with water-repellant rim.
5. Pre-incubate sections with TdT-buffer (*Appendix B.1.4*) — 5 min
6. Tap off TdT-buffer, do not wash; add TUNEL mixture (see below): spot 10–12 μl on coverslip (21 × 24 mm) and put on tissue section carefully — 60 min, 37°C
7. Wash with TB buffer (*Appendix B.1.5*) — 15 min (coverslips will drop off themselves)
8. Wash with TBS — 3 × 2 min
   Option: Heat-induced antigen retrieval with citrate buffer (*Appendix B.1.3* and *C.1.2*)
   wash with TBS
   Option: Store sections in TBS overnight at 4°C.
9. Normal goat serum, tap off do not wash — 10 min
10. Incubate with rabbit or mouse — overnight at 4°C or 60 min
    primary antibody
    wash with TBS — 3 × 2 min
11. EnVision+/HRP, GAR or Envision+/HRP, GAM + sheep anti-DIG/AP — 30 min
    wash with TBS — 3 × 2 min
12. Visualize AP activity with either NBT/BCIP substrate in blue/purple, or with Fast Red in red — 2–20 min
13. Wash with TBS — 3 × 2 min
14. Visualize HRP activity with DAB substrate* — 2–10 min
    Option: Blue nuclear counterstain with haematoxylin for the brown–red combination; Green nuclear counterstain with Methylgreen (*Appendix B.1.8*) for the brown–blue/purple combination.
15. Rinse with distilled water and coverslip aqueously with Faramount (DAKO).

---

*Development of HRP activity with DAB should be kept short, ensuring an yellowish/brown transparent reaction product.

Reagents	Code no.	Dilution
Proteinase K	Roche/Boehringer 745 723	
TdT-kit, 500 U	Roche/Boehringer 220 582	
dUTP-DIG, 25 nmol	Roche/Boehringer 1093 088	
Normal goat serum	DAKO X0907	1:10
EnVision™ + /HRP, goat anti-mouse Ig	DAKO K4000/4001	undiluted
EnVision™ + /HRP, goat anti-rabbit Ig	DAKO K4002/4003	undiluted
Sheep anti-DIG/AP	Roche/Boehringer 109 3274	1:200–1000*
NBT/BCIP substrate system	DAKO K0598	
Fast Red substrate system	DAKO K0597	

*For sections which underwent heat-induced antigen retrieval sheep anti-DIG/AP is used 1:200 diluted; without the tissue pretreatment sheep anti-DIG/AP is diluted up to 1:1000.

TUNEL mixture, e.g. for 8–9 tissue sections

Distilled water	68.0 µl
TdT-buffer 5× (from TdT kit)	20.0 µl
CoCl$_2$ (from TdT kit)	10.0 µl
dUTP-DIG, 1 nmol µl^{-1}	2.0 µl
TdT, 25 U µl^{-1} (from TdT kit)	0.4 µl
Total	100.0 µl

Appendix E

E.1 Non-commercial visualization systems

E.1.1 HRP

- *AEC*

  (Modified from Graham *et al.*, 1965.)

  (i)   25 mg 3-amino-9-ethylcarbazole (AEC) (Sigma A5754) dissolved in 2.5 ml N,N-dimethylformamide is mixed with 47.5 ml Na-acetate buffer (50 mM, pH 5.2);

  (ii)   filter the solution, and add 20 μl peroxide 30% just before use;

  (iii) stain for 5–20 min at room temperature;

  (iv)   results in a brick/red reaction product. Soluble in alcohols and xylene;

  (v)   note 1: this concentration of AEC chromogen (0.5 mg ml^{-1} end concentration) might be higher than most laboratories usually apply; a higher concentration of chromogen will result in a more intensely stained reaction product, but without causing non-specific background staining problems.

- *DAB*

  (Graham and Karnovski, 1966.)

  (i)   5 mg 3,3'-diaminobenzidin (DAB) (Sigma D5637) dissolved in 10 ml Tris–HCl buffer (50 mM, pH 7.8);

  (ii)   add 10 μl peroxide 30% just before use;

  (iii) stain for 5–20 min at room temperature;

  (iv)   results in a brown/yellow reaction product. Aqueously and organically insoluble;

  (v)   notes: colourshifting and intensification using metal ions is described by Hsu and Soban, 1982 (blue/black) and Nemes, 1987 (green).

- *TMB*

  (Modified from Buckel and Zehelein, 1981.)

(i)   dissolve 24 mg 3,3′,5,5′-tetramethylbenzidin (TMB) (Sigma T2885) and 80 mg dioctyl sodium sulphosuccinate (Sigma D0885) in 10 ml warm ethanol ($\approx$ 60°C). Mix with 30 ml citrate/phosphate buffer (100 mM, pH 5);

(ii)   add 10 µl peroxide 30% just before use;

(iii)   stain for 2–10 min at room temperature;

(iv)   results in a blue/green reaction product. Aqueously soluble;

(v)   note 1: after staining, wash briefly in distilled water and dry section completely. Soak in xylene and mount up organically.

(vi)   note 2: TMB chromogen results in a highly efficient/sensitive visualization of HRP activity. Therefore, it is recommended to have the primary antibody 10–50 times more diluted than with DAB or AEC as chromogen.

E.1.2 AP

- *Fast Blue / Fast Red*

  (Modified from Boorsma, 1984; Burnstone, 1961.)

(i)   20 mg naphthol-ASMX-phosphate (Sigma N4875) is dissolved in *N,N*-dimethylformamide and mixed with 100 ml Tris–HCl buffer (100 mM, pH 8.5); store 5 ml aliquots at −20°C;

(ii)   add to one 5 ml aliquot just before use: 1 mg Fast Blue BB (Sigma F3378) or 5 mg Fast Red TR (Sigma F1500);

(iii)   0.5 mM of levamisole (Sigma L9756) should be included for cryostat sections (2.5 µl of 1 M concentrate for 5 ml);

(iv)   stain for 5–20 min at room temperature;

(v)   results in a blue (Fast Blue BB) or red (Fast Red TR) reaction product. Soluble in alcohols and xylene;

(vi)   the reaction product obtained with Fast Red is also highly fluorescent (Speel *et al.*, 1992; Van der Loos and Becker, 1994);

(vii)   note 1: levamisole may also inhibit the marker alkaline phosphatase activity (Thisted, 1985), therefore do not include levamisole for paraffin sections;

(viii)   note 2: when phosphate-buffered saline is used throughout the incubation steps, the buffer salts should be washed out prior to the enzyme visualization using e.g. Tris–HCl buffer, because phosphate ions have an inhibitory effect on alkaline phosphatase activity.

NBT/BCIP

    (Modified from McGadey, 1970.)

(i) dissolve 18 mg 5-bromo-4-chloro-3-indolyl-phosphate (BCIP) (Roche/Boehringer 760 994) in 1 ml *N,N*-dimethylformamide; store at $-20°C$. Prepare nitro blue tetrazolium (NBT) (Sigma N6876) 5 mg ml^{-1} in distilled water (store at $-20°C$) and Tris–HCl buffer (100 mM, pH 9.5) including magnesium chloride (10 mg ml^{-1}).

(ii) mix 100 μl BCIP substrate solution with 9.34 ml Tris–HCl/Mg buffer and 0.66 ml NBT solution;

(iii) stain for 5–20 min at room temperature;

(iv) results in a purple/blue reaction product. Partly soluble in alcohols and xylene (see: description brown–purple/blue colour combination, *Section 5.3.2*);

(v) see also notes concerning levamisole and phosphate ions in the Fast Blue/Fast Red section.

- *New Fuchsin*

    (Modified from Feller *et al*., 1983.)

(i) dissolve 35 mg Naphthol-ASBI-phosphate (Sigma N2250) in 0.42 ml *N,N*-dimethylformamide and mix with 50 ml Tris–HCl buffer (50 mM, pH 9.7) including 378 mg 2-amino-2-methyl-1,3-propanediol (Sigma A9754) and 600 mg natrium chloride; store 5 ml aliquots at $-20°C$;

(ii) mix 83 μl freshly prepared New Fuchsin (Merck 4041) (12.5 mg 2 ml^{-1} 2 N HCl), with 52 μl sodium nitrite solution (Merck 6549) (20 mg ml^{-1} distilled water); stand for 2 min for diazonium reaction;

(iii) mix with 5 ml buffer/substrate solution; check pH to be 8.7;

(iv) stain for 5–20 min at room temperature;

(v) results in a red reaction product. Slightly soluble in alcohols (see description for red–turquoise colour combination);

(vi) see also notes concerning levamisole and phosphate ions in the Fast Blue/Fast Red section.

E.1.4.3 GAL

- *X-GAL*

    (Bondi *et al*., 1982.)

(i) dissolve 20 mg 5-bromo-4-chloro-3-indolyl-β-D-galactopyranoside (X-GAL) (Roche/Boehringer 651 745) in 1 ml *N,N*-dimethylformamide; store at $-20°C$. Prepare 50 ml of phosphate-buffered saline, pH 7.4, including 10 mg magnesium chloride.6H$_2$O, 49.5 mg potassium ferrihexacyanide and 63.5 mg potassium ferrohexacyanide; store at $+4°C$. Mix 10 μl of X-GAL substrate with 1 ml of iron phosphate solution;

(ii) stain for 10–60 min at 37°C;

(iii) results in a turquoise reaction product. Aqueously and organically insoluble.

- *Bluo-GAL*

    (Aguzzi and Theuring, 1994.)

(i) dissolve 20 mg 5-bromo-β-D-galactoside (Bluo-GAL) (Sigma B4387) in 1 ml *N,N*-dimethylformamide; store at $-20°C$. Mix 10 μl of X-GAL substrate with 1 ml of iron-phosphate solution (see X-GAL);

(ii) stain for 10–60 min at 37°C;

(iii) results in a blue reaction product. Aqueously and organically insoluble.

Appendix F

F.1 Commercial visualization systems

F.1.1 HRP

- *AEC*
  DAKO AEC substrate chromogen K3464, or for more sensitivity: DAKO AEC + substrate chromogen K3461 Note: the sensitivity/efficiency of DAKO K3461 is less than that of the procedure in *Appendix E.1.1.*
- *DAB*
  DAKO DAB liquid chromogen K3465, or for more sensitivity: DAKO DAB + liquid chromogen K3467.
- *TMB*
  True Blue™ peroxidase substrate, Kirkegaard & Perry Laboratories 71-00-64.

F.1.2 AP

- Fast Red
  DAKO Fast Red substrate system K0597 or K0699.
- New Fuchsin
  DAKO New Fuchsin substrate system K0596.
- NBT/BCIP
  DAKO NBT/BCIP substrate system K0598, Roche/Boehringer 1442 074.
- Vector Red
  Alkaline phosphatase kit I, Vector Laboratories SK-5100.

F.1.3 GAL

- *X-GAL*
  Biogenex detection kit HK162-5K.

F.1.4 Silver enhancement

- Silver enhancement
  R-Gent silver enhancement kit, Aurion 500.011.

Appendix G

G.1 Antibody conjugation protocols

G.1.1 Biotinylation or digoxigenin conjugation of antibodies

(See also Roche/Boehringer protocol with product info.)

- Dialyse protein solution (1–5 mg ml^{-1}) against PBS adjusted to pH 8.0, overnight at 4°C. Measure IgG concentration = $OD_{280} \times 1/1.4$.
- Dissolve N-hydroxysuccinimide ester in N,N-dimethylformamide (freshly prepared, do not store). Use either:
  D-biotinyl-ε-amidocaproic acid N-hydroxysuccinimide (Roche/Boehringer 1008 960)
  or
  digoxigenin 3-o-amidocaproic acid N-hydroxysuccinimide (Roche/Boehringer 1333 054).
- Add biotin- or DIG-ester to IgG under constant stirring (use glass tubes), see calculation below.
- Stir gently for 2 h at room temperature.
- Dialyse overnight at 4°C against TBS.
- Add BSA (1% final concentration) and sodium azide (0.1%).
- Store conjugate at 4°C.

Calculation example for the biotinylation of 400 μg IgG with a 10 × molar excess of biotin–ester:

400 μg/160 000 (= molecular mass IgG)	= 2.5×10^{-9} mol
10 × molar excess of biotin–ester	$\rightarrow 2.5 \times 10^{-8}$ mol
$2.5 \times 10^{-8} \times 454$ (= molecular mass biotin–ester)	= 1.14×10^{-5} g
	= 11.4 μg
Stock solution of biotin–ester: 2.2 mg ml^{-1}	$\rightarrow 5$ μl = 11.4 μg

Conjugates prepared with a 10 × molar excess of biotin- or DIG–ester, have been proven to yield satisfactory results in immunohistochemistry.

G.1.2 Conjugation with FITC

(The and Feldkamp, 1970.)

Reagents:
- purified immunoglobulin fraction
- fluorescein isothiocyanate (FITC) (Sigma F 7250)
- coupling buffer: sodium carbonate buffer (100 mM; pH 9.4) with 0.15 M NaCl.

Conjugation:
- dialyse protein solution (1–5 mg ml^{-1}) against coupling buffer, overnight at 4°C
- dissolve FITC in dimethylsulphoxide (1 mg ml^{-1} in glass tube)
- add FITC slowly to the protein solution under gentle stirring ratio: 15 µl FITC mg protein^{-1}
- stir for 2 h at room temperature in the dark
- bring mixture onto PD-10 G-25 (Amersham Pharmacia Biotech) column equilibrated with phosphate-buffered saline
- collect the first 'yellow peak'
- dialyse against phosphate-buffered saline overnight at 4°C
- Measure OD_{280} and OD_{495} for F/P ratio
- add BSA (1% final concentration) and sodium azide (0.1%)
- store conjugate at 4°C.

Protein concentration in IgG–FITC conjugate (mg ml^{-1}) =

$$\frac{[OD_{280} - (0.35 \times OD_{495})] \times \text{sample dilution}}{1.4}.$$

$$\text{F/P ratio} = \frac{OD_{495}/68\,000}{\text{protein conc. (mg ml}^{-1})/150\,000}$$

For immunofluorescence: F/P should be >1.5 and < 3.0.
For FACS and double staining via anti-FITC: F/P > 5.0.

G.1.3 Conjugation with β-galactosidase

Because of the poor availability of directly GAL-conjugated antibodies, this protocol according to Ishikawa *et al.* (1983) is given here.

- Add to a solution of 1.4 mg IgG (= 9.3 nmol) in 0.5 ml Na-phosphate buffer (100 mM, pH 7.0), 50 μl 3-maleimidobenzoyl-*N*-hydroxysuccinimide ester (MBS-ester) (Pierce) (0.9 mg ml^{-1} *N,N*-dimethylformamide).
- Incubate the reaction mixture for 30 min at 30°C.
- Pass it over a Sephadex G-25 (Amersham Pharmacia Biotech) column (1.0 × 40 cm), equilibrated with Na-phosphate buffer (100 mM, pH 6.5). Collect the first peak containing IgG-maleimide.
- For conjugation, mix the pooled maleimide-IgG (= 8.3 nmol) with 1.5 mg (= 2.8 nmol) of lyophilized β-galactosidase from *E. coli* (Pierce).
- Incubate the reaction mixture for 20 h at 4°C.
- Pass it over a Sephacryl S-400 (Amersham Pharmacia Biotech) column (1.6 × 50 cm), equilibrated with Na-phosphate (10 mM, pH 6.5) with 100 mM NaCl, 1 mM MgCl$_2$. Measure OD$_{280}$.
- Test β-galactosidase activity in the fractions: mix 150 μl β-galactosidase visualization reagent (*Appendix E.1.3*) with a 20 μl sample from a fraction. Incubate for 15 min at 37°C and read peak fractions.
- IgG-β-galactosidase conjugate will be found in the first peak. Pool and concentrate to ±0.5 ml.
- Add BSA (1% final concentration) and sodium azide (0.1%).
- Store conjugate at 4°C.

Appendix H

H.1 Miscellaneous

H.1.1 How to get started with double immunohistochemistry?

Theory remains theory until the moment that it has been proven to work. This statement definitely accounts for the procedures in this handbook. The following 10 steps may help to get started with immunoenzyme double staining techniques.

1. Read this handbook carefully.
2. Cut (semi) serial sections and perform single immuno staining; keep the rest of the sections in stock for double staining
3. Determine the characteristics of primary antibodies (species, Ig isotype, IgG subclass), and check for the possibility of commercially available conjugates (biotin, FITC, phycoerythrin, DAKO EPOS™/HRP)
4. Use the flowchart (*Table 5.1*) to find the simplest and most suitable double-staining method.
5. When finishing with two mouse monoclonal antibodies of similar Ig isotype or subclass: search for one alternative primary antibody with different characteristics or labels. In case of expected co-localization, use the DAKO ARK kit for biotinylation of one primary antibody (see: *Section 5.1*). The *sequential* double staining procedure should be applied as last resort (see: *Section 3.1* and *Appendix D.1.1*).
6. Select a colour combination. Try to imagine how your double staining would look in a certain colour combination during observation of your single staining specimens. Most of the time your imagination will be similar with the recommendations given in *Table 6.1*. The expected double-staining pattern will have to meet the colour combination characteristics (*Section 5.3*).
7. Adapt the dilution of the primary antibody(ies) to the sensitivity/efficiency of the applied detection- and chromogen system (*Tables 5.2 and 5.3*).

8. Design a full double-staining method on paper including a schematic drawing; check your design for possible interspecies cross-reactivity.

9. Perform your double staining and compare with the original single-staining specimens. Do not handle too many slides in one experiment, especially not when testing more than one primary antibody combination. Include a control experiment: two half-double-staining experiments, with the omission of one or both primary antibodies (*Section 4.2*).

10. If necessary: perform the double staining in a reversed colour combination.

H.1.2 Answers to trouble shooting Nos 1–6

No. 1. The swine anti-rabbit/AP reagent from the second step cocktail, will react *in vitro* with the rabbit anti-mouse Ig/HRP conjugate. Therefore, the mouse antibody will be finally labelled with both enzymatic tracers.

Action: Replace the reagents from the second step cocktail with reagents raised in the same host, e.g. goat anti-mouse Ig/HRP + goat anti-rabbit Ig/AP.

No. 2. Commercially available second-step reagents designed for use on human tissue specimens, are usually not absorbed for rat Igs. The closer the mammalian species are related evolutionary, the stronger the similarity of their Ig structures, and therefore the stronger the cross-reactions will be. In this case the anti-mouse Ig reagent will react strongly with the rat Ig endogenously present in tissue section.

Action: Mix 10–20% normal rat serum, containing normal rat Ig, with the second-step reagent cocktail 30 min before use. Those anti-mouse Igs which might react with rat Ig, will now be bound *in vitro* with the normal rat Ig, and cannot bind to the rat Ig in the tissue section anymore.

No. 3. The goat anti-rabbit Ig/HRP reagent will react with the rabbit anti-mouse Ig/biotin step.

Action: Replace rabbit anti-mouse Ig/biotin with goat anti-mouse Ig/biotin.

No. 4. Concentrated normal goat serum was used for blocking non-specific binding. This reagent will stick to tissue elements non-specifically. As a consequence, the swine anti-goat Ig/HRP reagent will certainly react with it, and will result in a heavy background staining.

Action: Replace normal goat serum with normal swine serum.

No. 5. In this protocol there are two mistakes:

(1) Normal goat serum used for blocking non-specific binding will be detected by the swine anti-goat Ig/HRP reagent (see also Trouble shooting No. 4).

Action: Replace normal goat serum with normal swine serum.

(2) Goat anti-rabbit Ig/AP will be captured by the swine anti-goat Ig/HRP reagent.

Action: Either replace goat anti-rabbit Ig/AP with swine anti-rabbit Ig/AP, or insert a normal goat serum step before adding the goat anti-rabbit Ig/AP reagent. For time saving reasons normal goat serum should be mixed with the normal mouse serum step.

No. 6. In the streptavidin-biotin complex with HRP, the HRP enzyme is biotinylated. Therefore, the streptavidin/AP reagent will react with these free biotin groups.

Action: Insert step with 0.1% (strep)avidin before the d-biotin step, as is regularly done with blocking endogenous biotin (see *Section 5.2*).

Although it seems feasible in this protocol to apply a streptavidin-biotin detection twice, this double-staining procedure is not encouraged. For optimal performance of the unlabelled antibody a detection procedure with e.g. DAKO EnVision™ is recommended.

H.1.3 Detailed immunohistochemistry working protocols for the colour plates

The following detailed protocols have been applied for the colour plates 1–15. TBS washing steps have been omitted from the protocols. Incubation steps are performed at room temperature if no temperature is indicated. Vendor and ordering details of the detection reagents can be found with the indicated protocol in the Appendix. With some protocols the figure number in *Chapter 3* of a schematic drawing of the staining procedure is also indicated. With *Plate 8 A–M* a selection is made for G/H and I/J, illustrating the adaption of primary antibody dilution with respect to the chromogen used.

Plate 1: CD79a / CD3 (aorta aneurism, paraffin section)
Indirect/indirect double staining procedure.
See: *Appendix D.1.3, and Figure 3.3*

Dewaxing and rehydration, heat-induced antigen retrieval with citrate pH 6.0 (15 min, 100°C), blocking of endogeneous peroxidase activity with peroxide (0.3%) in methanol (20 min), washing buffer.

NGS	1:10		15 min
⌈Anti-CD79a, clone JCB117, mouse (DAKO)	1:20		
		(cocktail)	60 min
⌊Anti-CD3, rabbit (DAKO)	1:200		
⌈GAM/AP	1:20		
		(cocktail)	30 min
⌊GAR/HRP	1:50		
Visualization of AP activity in blue			
(Fast Blue BB)			20 min
Visualization of HRP activity in brown			
(DAKO DAB+)			5 min

Plate 2: Vimentin/Cytokeratin (kidney adenocarcinoma, cryostat section)
Indirect/direct double staining procedure.
See: *Appendix D.1.5*, and *Figure 3.6*

Blocking of endogeneous peroxidase activity with peroxide (0.3%) and sodium azide (0.1%) in TBS washing buffer (20 min), washing buffer.

Normal goat serum	1:10		15 min
Anti-Vimentin, clone V9,			
mouse (DAKO)	1:40		60 min
GAM/biotin	1:200		30 min
⎰Streptavidin/AP	1:100		
⎱		(cocktail)	30 min
Normal mouse serum	1:10		
Anti-Cytokeratin/FITC, clone MNF116,			
mouse (DAKO)	1:50		60 min
Rabbit anti-FITC	1:1000		15 min
GAR/HRP	1:50		30 min
Visualization of AP activity in blue			
(Fast Blue BB)			20 min
Visualization of HRP activity in red			
(AEC)			10 min

Plate 3: Ki67/Cytokeratin (mammary carcinoma, paraffin section)
Indirect/indirect double staining procedure.
See: *Appendix D.1.3*, and *Figure 3.4*

Dewaxing and rehydration, heat-induced antigen retrieval with citrate pH 6.0 (15 min, 100°C), blocking of endogeneous peroxidase activity with peroxide (0.3%) in methanol (20 min), washing buffer.

NGS	1:10		15 min
⎰Anti-Cytokeratin, clone MNF116,			
⎱ mouse (DAKO)	1:10		
		(cocktail)	60 min
Anti-Ki67, rabbit (DAKO)	1:50		
⎰GAM/biotin	1:200		
⎱		(cocktail)	30 min
EnVision+™/HRP, GAR	undil		
Strep/AP	1:100		30 min
Visualization of AP activity in blue			
(Fast Blue BB)			20 min
Visualization of HRP activity in red			
(AEC)			10 min

Plate 4: p53/Cytokeratin (colon carcinoma, paraffin section)

Sequential double staining procedure.
See: *Appendix D.1.1*, and *Figure 3.1*

Dewaxing and rehydration, heat-induced antigen retrieval with citrate pH 6.0 (15 min, 100°C), blocking of endogeneous peroxidase activity with peroxide (0.3%) in methanol (20 min), washing buffer.

NGS	1:10	15 min
Anti-Cytokeratin, clones AE1+AE3, mouse		
(DAKO)	1:50	30 min
EnVision+™/HRP, GAM+GAR	undiluted	30 min
Visualization of HRP activity in brown		
(DAKO DAB+)		10 min
Double staining blocking solution		30 min
Anti-p53, clone DO-7, mouse (DAKO)	1:50	60 min
EnVision™/AP, GAM+GAR	undiluted	60 min
Visualization of AP activity in red		
(DAKO Fast Red)		7 min
Nuclear counterstain in blue with haematoxylin		

Plate 5: CD3/CD25, IL-2 Receptor (tonsil, cryostat section)

Direct/direct double staining procedure.
See: *Appendix D.1.2*, and *Figure 3.2*

Blocking of endogeneous peroxidase activity with peroxide (0.3%) and sodium azide (0.1%) in TBS washing buffer (20 min), washing buffer.

Normal swine serum	1:10		15 min
⎰Anti-CD3/PE, clone SK7, mouse (BD)	1:20		
⎱		(cocktail)	60 min
Anti-CD25/FITC, clone 2A3, mouse (BD)	1:50		
⎰Rabbit anti-PE	1:200		
⎱		(cocktail)	15 min
Goat anti-FITC	1:400		
⎰SAR/AP	1:10		
⎱		(cocktail)	30 min
SAG/HRP	1:50		
Visualization of AP activity in blue			
(Fast Blue BB)			15 min
Visualization of HRP activity in red			
(AEC)			10 min

Plate 6: CD34/von Willebrand factor (tonsil, cryostat section)

Indirect/indirect double staining procedure.
See: *Appendix D.1.3*, and *Figure 3.3*

Blocking of endogeneous peroxidase activity with peroxide (0.3%) and sodium azide (0.1%) in TBS washing buffer (20 min), washing buffer.

NGS	1:10		15 min
Anti-CD34, clone QBEnd10, mouse (DAKO)	1:200		
		(cocktail)	60 min
Anti-von Willebrand factor, rabbit (DAKO)	1:10 000		
GAM/AP	1:20		
		(cocktail)	30 min
EnVision+™/HRP, GAR	undil		
Visualization of AP activity in purple/blue			
(NBT/BCIP)			10 min
Visualization of HRP activity in brown			
(DAKO DAB+)			2 min
Nuclear counterstain in green with Methylgreen			

Plate 7: α-actin/CD68 (atherosclerotic lesion in carotis, paraffin section)

Indirect/indirect double staining procedure.
See: *Appendix D.1.4*, and *Figure 3.5*

Dewaxing and rehydration, heat-induced antigen retrieval with citrate pH 6.0 (15 min, 100°C), washing buffer.

NGS	1:10		15 min	
Anti-α-actin, clone 1A4, mouse IgG2a				
(DAKO)	1:50			
		(cocktail)	60 min	
Anti-CD68, clone PG-M1, mouse IgG3				
(DAKO)	1:50			
GAM IgG2a/biotin	1:100			
		(cocktail)	30 min	
GAM IgG3/AP	1:20			
Strep/GAL	1:40		30 min	
Visualization of GAL activity in turquoise				
(X-gal/ferri-ferrocyanide)			30 min	37°C
Visualization of AP activity in red				
(DAKO New Fuchsin)			5 min	

Plate 8G and H: CD4/CD8 (thymus, cryostat section)

Indirect/direct double staining procedure.
See: *Appendix D.1.5*, and *Figure 3.6*

Blocking of endogeneous peroxidase activity with peroxide (0.3%) and sodium azide (0.1%) in TBS washing buffer (20 min), washing buffer.

G			**H**		
Normal goat serum	1:10	15 min	Normal goat serum	1:10	15 min
Anti-CD4, MT310,	1:20	60 min	Anti-CD4, MT310,	1:50	60 min
mouse (DAKO)			mouse (DAKO)		
GAM/biotin	1:200	30 min	GAM/biotin	1:200	30 min
⎧Strep/AP	1:100		⎧Strep/HRP	1:400	
⎨	(cocktail)	30 min	⎨	(cocktail)	30 min
⎩Normal mouse serum	1:10		⎩Normal mouse serum	1:10	
Anti-CD8/FITC, DK25,			Anti-CD8/FITC, DK25,		
mouse (DAKO)	1:200	60 min	mouse (DAKO)	1:100	60 min
Rabbit anti-FITC	1:1000	15 min	Rabbit anti-FITC	1:1000	15 min
GAR/HRP	1:50	30 min	GAR/AP	1:20	30 min
Visualization of AP			Visualization of AP		
activity in blue			activity in blue		
(Fast Blue)		15 min	(Fast Blue)		15 min
Visualization of HRP			Visualization of HRP		
activity in red			activity in red		
(AEC)		10 min	(AEC)		10 min

Plate 8I and J: CD4/CD8 (thymus, cryostat section)

Indirect/direct double staining procedure.
See: *Appendix D.1.5*, and *Figure 3.6*

Washing buffer.

I			**J**		
Normal goat serum	1:10	15 min	Normal goat serum	1:10	15 min
Anti-CD4, MT310,		60 min	Anti-CD4, MT310,		
mouse (DAKO)	1:10		mouse (DAKO)	1:50	60 min
GAM/biotin	1:200	30 min	GAM/biotin	1:200	30 min
⎧Strep/GAL	1:40		⎧Strep/AP	1:100	
⎨	(cocktail)	30 min	⎨	(cocktail)	30 min
⎩Normal mouse serum	1:10		⎩Normal mouse serum	1:10	
Anti-CD8/FITC, DK25,			Anti-CD8/FITC, DK25,		
mouse (DAKO)	1:20	60 min	mouse (DAKO)	1:50	60 min
Rabbit anti-FITC	1:1000	15 min	Rabbit anti-FITC	1:1000	15 min
GAR/AP	1:20	30 min	GAR/GAL	1:10	30 min
Visualization of GAL			Visualization of GAL		
activity in turquoise			activity in turquoise		
(X-gal)		15 min	(X-gal)		15 min
Visualization of AP			Visualization of HRP		
activity in red			activity in red		
(Fast Red)		10 min	(Fast Red)		10 min

Plate 9: CD3/HLA-DR (atherosclerotic coronary artery, cryostat section)

Indirect/direct double staining procedure.
See: *Appendix D.1.5*, and *Figure 3.6*

Blocking of endogeneous peroxidase activity with peroxide (0.3%) and sodium azide (0.1%) in TBS washing buffer (20 min), washing buffer.

A			**B**		
Normal goat serum	1:10	15 min	Normal goat serum	1:10	15 min
Anti-HLA-DR, CR3/34, mouse (DAKO)	1:500	60 min	Anti-HLA-DR, CR3/34, mouse (DAKO)	1:200	60 min
GAM/biotin	1:200	30 min	GAM/biotin	1:200	30 min
Strep/HRP	1:400		Strep/AP	1:100	
	(cocktail)	30 min		(cocktail)	30 min
Normal mouse serum	1:10		Normal mouse serum	1:10	
Anti-CD3/FITC, SK7, mouse (BD)	1:100	60 min	Anti-CD8/FITC, SK7, mouse (BD)	1:200	60 min
Rabbit anti-FITC	1:1000	15 min	Rabbit anti-FITC	1:1000	15 min
GAR/AP	1:20	30 min	GAR/HRP	1:50	30 min
Visualization of AP activity in blue (Fast Blue)		15 min	Visualization of AP activity in blue (Fast Blue)		15 min
Visualization of HRP activity in red (AEC)		10 min	Visualization of HRP activity in red (AEC)		10 min

Plate 10: IL-1α/α-actin (atherosclerotic aorta segment, cryostat section)

Indirect/indirect double staining procedure
See: *Appendix D.1.5, D.1.6*

Washing buffer.

NGS	1:10		15 min
Anti-Interleukin 1α, rabbit (Genzyme/R&D Systems)	1:200		overnight 4°C
Anti-α-actin, clone 1A4, mouse (DAKO)	1:50		60 min
GAR/biotin	1:400		
		(cocktail)	30 min
GAM/AP	1:50		
IGS blocking			15 min
Goat anti-biotin/Ultra small gold	1:40		150 min
Visualization of AP activity in red (Vector Red)			5 min
Fixation with buffered formalin			10 min
Wash with distilled water			5 min
Silver enhancement (Aurion R-Gent)			15 min
Nuclear counterstain in blue with haematoxylin			

Plate 11: TUNEL/CD20, B-cells (tonsil, paraffin section)
See: *Appendix D.1.7*

Performance of TUNEL according to protocol in *Appendix D.1.7*, steps 1–8

NGS	1:10	15 min
Anti-CD20, clone L26, mouse (DAKO)	1:50	60 min
EnVision+™/HRP, GAM	undiluted	30 min
Visualization of AP activity in purple/blue		
(DAKO NBT/BCIP)		10 min
Visualization of HRP activity in brown		
(DAKO DAB+)		10 min
Nuclear counterstain in green with Methylgreen		

Plate 12: TUNEL/CD68, macrophages (tonsil, paraffin section)
See: *Appendix D.1.7*

Performance of TUNEL according to protocol in *Appendix D.1.7*, steps 1–8. Heat-induced antigen retrieval with citrate buffer pH 6.0 (15 min, 100 C).

NGS	1:10	15 min
Anti-CD68, clone PG-M1, mouse (DAKO)	1:500	60 min
EnVision+™/HRP, GAM	undiluted	30 min
Visualization of AP activity in red		
(DAKO Fast Red)		10 min
Visualization of HRP activity in brown		
(DAKO DAB+)		10 min
Nuclear counterstain in blue with haematoxylin		

Plate 13: α-actin/collagen type III/von Willebrand factor (myocardium, cryostat section)

Indirect/indirect/indirect triple staining procedure.

Blocking of endogeneous peroxidase activity with peroxidase (0.3%) and sodium azide (0.1%) in TBS washing buffer (20 min), washing buffer.

NGS	1:10		15 min	
⎧ Anti-α-actin, 1A4,				
⎪ mouse IgG2a (DAKO)	1:100			
⎨ Anti-collagen type III, HWD1.1,				
⎪ mouse IgG1 (Biogenex)	1:4000	(cocktail)	overnight	4°C
⎪ Anti-von Willebrand factor,				
⎩ rabbit (DAKO)	1:10 000			
⎧ GAM-IgG2a/biotin	1:100			
⎨ GAM-IgG1/AP	1:20	(cocktail)	30 min	
⎩ GAR/HRP	1:50			
Streptavidin/GAL	1:40		30 min	
Visualization of GAL activity in turquoise				
(X-GAL/ferriferrocyanide)			20 min	37°C
Visualization of AP activity in blue				
(Fast Blue BB)			5 min	
Visualization of HRP activity in red (AEC)			10 min	

Plate 14: α-actin/CD68/CD3 (atherosclerotic aorta segment, cryostat section)

Indirect / indirect / direct triple staining procedure.

Blocking of endogeneous peroxidase activity with peroxide (0.3%) and sodium azide (0.1%) in TBS washing buffer (20 min), washing buffer.

NGS	1:10		15 min
⎰Anti-α-actin, clone 1A4, mouse IgG2a (DAKO)	1:50		
		(cocktail)	60 min
⎱Anti-CD68, clone EBM11, mouse IgG1 (DAKO)	1:50		
⎰GAM-IgG2a/biotin	1:100		
		(cocktail)	30 min
⎱GAM-IgG1/AP	1:20		
⎰Normal mouse serum	1:10		
		(cocktail)	30 min
⎱Strep/GAL	1:40		
Anti-CD3/FITC, clone SK7, mouse IgG1 (BD)	1:200		60 min
Rabbit anti-FITC	1:1000		15 min
GAR/HRP	1:50		30 min
Visualization of GAL activity in turquoise (X-GAL/ferri-ferrocyanide)			30 min 37°C
Visualization of AP activity in blue (Fast Blue BB)			10 min
Visualization of HRP activity in red (AEC)			10 min

Plate 15: MUC5A/MUC6/*Helicobacter pylori* (stomach, paraffin section)

Sequential/indirect/indirect triple staining procedure.

Dewaxing and rehydration, heat-induced antigen retrieval with citrate pH 6.0 (15 min, 100°C), blocking of endogeneous peroxidase activity with peroxide (0.3%) in methanol (20 min), washing buffer.

NGS	1:10	15 min	
Anti-secretory mucin type 6, rabbit (gift)	1:200	overnight	4°C
GAR/HRP	1:50	30 min	
Visualization of HRP activity in brown (DAB)		10 min	
Heat-induced antigen retrieval with citrate pH 6.0		15 min	100°C
⎰Anti-secretory mucin type 5AC, mouse (Novocastra)	1:50		
⎱	(cocktail)	overnight	4°C
Anti-*Helicobacter pylori*, rabbit (DAKO)	1:5000		
⎰GAM/biotin	1:200		
⎱	(cocktail)	30 min	
GAR/AP	1:20		
Streptavidin/GAL	1:40	30 min	
Visualization of GAL activity in turquoise (X-GAL/ferri-ferrocyanide)		20 min	37°C
Visualization of AP activity in red (DAKO Fast Red)		7 min	
Nuclear counterstain in blue with haematoxylin			

Appendix I

I.1 Reagent sources

Aurion BV
Costerweg 5, NL-6702 AA Wageningen, The Netherlands
Tel: +31 8370 97676; Fax: +31 8370 15955; website: http://www.aurion.nl; e-mail: leunissen@aurion.nl

Amersham Pharmacia Biotech
Björkgatan 39, 751 84 Uppsala, Sweden
Tel: +46 18 165000; Fax: +46 18 166458; website: http://www.apbiotech.com

Becton Dickinson Immunocytometry Systems
2350 Qume Drive, San Jose, CA 95131-1807, USA
Tel: +1 800 448 2347; Fax: +1 408 954 2347; website: http://www.bdfacs.com

Biogenesis Ltd.
7 New Fields, Stinford Road, Poole BH17 0NF, UK
Tel: +44 1202 660006; Fax: +44 1202 660020;
website: http://www.biogenesis.co.uk/home/; e-mail: biogenesis@lds.co.uk

BioGenex Laboratories
4600 Norris Canyon Road, San Ramon, CA 94583,USA
Tel: +1 925 275 0550; Fax: +1 925 275 0580;
website: http://www.biogenex.com; e-mail: sales@biogenex,com

DAKO A/S
Produktionsvej 42, DK-2600 Glostrup, Denmark
Tel: +45 44 859500; Fax: +45 44 859595; website: http://www.dako.com

DAKO Corporation
6392 Via Real, Carpinteria, CA 93013, USA
Tel: +1 805 566 6655; Fax: +1 805 566 6688; website: http://www.dakousa.com

Genzyme see: R & D Systems Europe, Ltd
4–10 The Quadrant, Barton Lane, Abingdon, Oxon OX14 3YS, UK
Tel: +44 1235 551100; Fax: +44 1235 533420;
website: http://www.rndsystems.com;
e-mail: USA and Canada: infor@mdsystems.com;
e-mail Europ: infor@rndsystems.co.uk

Kirkegaard & Perry Laboratories
2 Cessna Court, Gaithersburg, MD 20879-47174, USA
Tel: +1 301 948 7755; Fax: +1 301 948 0169; website: http://www.kpl.com;
e-mail: reagents@kpl.com

Novocastra Laboratories Ltd.
Balliol Business Park West, Benton Lane, Newcastle upon Tyne
NE12 8EW, UK
Tel: +44 191 2150567; Fax: +44 191 1152;
website: http://www.novocastra.co.uk

Pierce
P.O. Box 117, Rockford, IL 61105, USA
Tel: +1 815 968 0747; Fax: +1 815 968 7316;
website: http://www.piercenet.com

Roche Diagnostics/Boehringer Mannheim, GmbH
Sandhofer Strasse 116, 68305 Mannheim, Germany
Tel: +49 621 759 8568; Fax: +49 621 759 4083;
website: http://biochem.roche

Southern Biotechnology Associates Inc.
PO Box 26221, Birmingham, AL 35226, USA
Tel: +1 205 945 1774; Fax: +1 205 945 8768;
website: http://www.southernbiotech.com;
e-mail: techserv@SouthernBiotech.com

Vector Laboratories Inc.
30 Ingold Road, Burlingame, CA 94010, USA
Tel: +1 650 697 3600; Fax: +1 650 697 0339;
website: http://www.vectorlabs.com; e-mail: vector@vectorlabs.com

Appendix J

J.1 Immunochemistry Web Sites

The following internet sites may provide you with additional information on immunohistochemistry.

http://immuno.hypermart.net
This site contains a comprehensive list of immunohistochemical resources on the internet including the "Immuno Forum" for questions and answers on detailed immunohistochemical problems.

http://www.immunoquery.com
To enter this site one has to be a subscriber. The site provides the staining patterns of numerous antibodies for many diseases, and *vice versa*, diseases linked with antibody panels.

http://www.jhc.org
This site is the online version of the *Journal of Histochemistry and Cytochemistry*. Abstracts only are available from January 1975. Full text and abstracts are available from January 1997.

http://www.nsh.org
This site is run by the National Society for Histotechnology in the USA. It provides information on forthcoming meetings, educational activities, the *Journal of Histotechnology* and links to the histochemical societies of the individual States.

http://www.rms.org.uk
This site is run by the Royal Microscopical Society in the United Kingdom. It provides information on forthcoming meetings, educational activities, the *Journal of Microscopy*, the Proceedings of the Royal Microscopical Society, the series of technical handbooks and links to other similar microscopical societies around the world.

http://link.springer.de/link/journals/00418/index.htm
This site is the online version of the journal *Histochemistry and Cell Biology*. Abstracts only are available from January 1994. Full text and abstracts are available from January 1996.

http://www.serotec.co.uk/anti.html
This Serotec Ltd. site offers the opportunity to search for any desired antibody (free of charge).

http://www.serotec.co.uk/board/index.html
This Serotec Ltd. site is called the "Antibody discussion group". Like the previously mentioned "Immuno Forum" it offers the possibility for questions and answers on immunohistochemical problems and the ability to search for rare "home-made" antibodies.

Index